Lecture Notes in Mathematics

Edited by A. Dold and B. Eckmann

1119

Recent Mathematical Methods in Dynamic Programming

Proceedings of the Conference held in Rome, Italy,
March 26–28, 1984

Edited by I. Capuzzo Dolcetta, W. H. Fleming and T. Zolezzi

Springer-Verlag
Berlin Heidelberg New York Tokyo

Editors

Italo Capuzzo Dolcetta
Dipartimento di Matematica "G. Castelnuovo"
Università di Roma "La Sapienza"
Città Universitaria, 00185 Rome, Italy

Wendell H. Fleming
Lefschetz Center for Dynamical Systems
Division of Applied Mathematics, Brown University
Providence, Rhode Island 02912, USA

Tullio Zolezzi
Università di Genova
Istituto di Matematica, Via L. B. Alberti, 4
16132 Genova, Italy

AMS Subject Classification (1980): 49C

ISBN 3-540-15217-2 Springer-Verlag Berlin Heidelberg New York Tokyo
ISBN 0-387-15217-2 Springer-Verlag New York Heidelberg Berlin Tokyo

Printed in Germany

Printing and binding: Beltz Offsetdruck, Hemsbach / Bergstr.
2146/3140-543210

PREFACE

This volume contains contributions by speakers at an international conference on Recent Mathematical Methods in Dynamic Programming, held at the Università di Roma "La Sapienza", March 26-28, 1984. The ten lectures presented dealt with analytical, numerical, and applied aspects of recent research in control, as well as related mathematical questions concerning partial differential equations, functional analysis, and stochastic processes. On the analytical side the following topics were included: (1) stochastic control for Markov diffusions and for processes with jumps, impulsive control, connections with large deviations theory and singular perturbations; (2) PDE - viscosity solution methods for deterministic and stochastic control problems; (3) deterministic control, necessary and sufficient dynamic programming conditions for optimality; (4) infinite dimensional state space problems, dynamical programming and variational inequality methods. The lectures presented by J-P. Quadrat and E. Rofman reported work at INRIA on advanced numerical techniques for feedback control problems, with application to energy production systems.

We wish to thank Dipartimento di Matematica - Istituto "G.Castelnuovo" of the University of Rome "La Sapienza" for support and organization, and in particular M. Falcone for valuable help in organizing the meeting.

We thank Ministero della Pubblica Istruzione (progetto 40%, "Calcolo delle variazioni") and Comitato per la Matematica del C.N.R. for financial support. In this respect we like to thank professors L. Amerio and M. Biroli for their cooperation.

I. Capuzzo Dolcetta

W.H. Fleming

T. Zolezzi

TABLE OF CONTENTS

ADDRESSES OF THE AUTHORS.

V.BARBU. University of Iasi. Iasi 6600 , Romania.

A. BENSOUSSAN. INRIA. Domaine de Voluceau. B.P. 105 . Rocquencourt. 78153 Le Chesnay Cedex. France.

G.DA PRATO. Scuola Normale Superiore. 56100 Pisa , Italy.

W. H. FLEMING. Lefschetz Center for Dynamical Systems. Division of Applied Mathematics. Brown University. Providence, Rhode Island 02912 , U.S.A.

C.GOMEZ , J.-P. QUADRAT , A.SULEM . INRIA. Domaine de Voluceau. B.P. 105 . Rocquencourt. 78153 Le Chesnay Cedex, France.

P.L.LIONS. CEREMADE. Université Paris-Dauphine. Place de Lattre de Tassigny . 75775 Paris Cedex 16 , France.

J.L.MENALDI . Department of Mathematics , Wayne State University. Detroit, MI 48202, U.S.A.

M. ROBIN. INRIA. Domaine de Voluceau. B.P.105 . Rocquencourt. 78153 Le Chesnay Cedex, France.

U. MOSCO . Dipartimento di Matematica "G. Castelnuovo". Università di Roma La Sapienza. Città Universitaria. 00185 Roma , Italy.

E.ROFMAN. INRIA. Domaine de Voluceau. B.P. 105 . Rocquencourt . 78153 Le Chesnay Cedex, France.

R.VINTER. Department of Electrical Engineering. Imperial College of Science and Technology. London SW7 2BT , United Kingdom.

THE TIME OPTIMAL CONTROL OF VARIATIONAL INEQUALITIES. DYNAMIC PROGRAMMING AND THE MAXIMUM PRINCIPLE.

Viorel Barbu
University of Iasi
Iasi 6600, Romania

1. INTRODUCTION

We are concerned here with the nonlinear control process

(1.1) $y'(t) + Ay(t) + Fy(t) \ni u(t)$ a.e. $t > 0$

$y(o) = y_0$

in the space $H = L^2(\Omega)$ (Ω is a bounded and open subset of R^N). Here $A: D(A) \subset H \longrightarrow H$ is a linear self adjoint operator which satisfies the coercivity condition

(1.2) $(Ay,y) \geqslant \omega |y|_2^2 \quad \forall\ y \in D(A)$

for some $\omega > 0$. The operator $F: H \longrightarrow H$ is defined by

(1.3) $(Fy) = \{w \in L^2(\Omega); w(x) \in \beta(y(x))$ a.e. $x \in \Omega\}$

where β is a maximal monotone graph in $R \times R$ such that $0 \in \beta(0)$. Further, we shall assume that

(1.4) $(Ay, \beta_\lambda(y)) \geqslant -C_1|y|_2^2 - C_2 \quad \forall\ y \in D(A), \ \lambda > 0$

where $\beta_\lambda = \lambda^{-1}(1-(1+\lambda\beta)^{-1})$. In the sequel we will denote by $(\cdot,\cdot)$ the usual scalar product in $H = L^2(\Omega)$ and by $|\cdot|_2$ the corresponding norm.

It is well known (see for instance [6], [7], [9]) that under the above assumptions, $A+F = \partial\varphi$ where $\partial\varphi: H \longrightarrow H$ is the subdifferential of the lower semicontinuous convex function

(1.5) $\varphi(y) = \frac{1}{2}(Ay,y) + \int_\Omega j(y)dx \quad , \quad y \in D(A^{1/2})$

where $\beta = \partial j$.

Throughout in the following we shall denote by V the space $D(A^{1/2})$ endowed with the graph norm $\| y\| = |A^{1/2}|_2$ and assume that the injection of V into H is compact. In particular, this implies that every level subset $\{y \in H;\ \varphi(y) \leq \lambda\}$ is compact.

According to standard existence results for evolution equations of gradient type, see ([2], [9]), for every $y_0 \in \overline{D(\varphi)}$, $T > 0$ and $u \in L^2(0,T;H)$ the Cauchy problem (1.1) admits a unique solution $y = y(t,y_0,u) \in C([0,T];H) \cap W^{1,2}([\delta,T];H) \cap L^2(\delta;T;D(A))$ for every

$\delta > 0$. If $y_0 \in D(\varphi)$ then $y(t,y_0,u) \in W^{1,2}([o,T];H) \cap L^2(o,T;D(A))$. Here $W^{1,2}([\delta,T];H)$ is the space $\{y \in L^2(\delta,T;H); y' \in L^2(\delta,T;H)\}$ and y' is the strong derivative of y.

A typical example is the nonlinear heat equation

$$(1.6) \quad \frac{\partial y}{\partial t} - \Delta y + \beta(y) \ni u \quad \text{in } \Omega \times R^+$$

$$y(o) = y_0$$

with Dirichlet or Neumann homogeneous boundary conditions. In this case $A = -\Delta$ with $D(A) = H_0^1(\Omega) \cap H^2(\Omega)$ or $D(A) = \{y \in H^2(\Omega); \frac{\partial y}{\partial \nu} + \alpha y = o \text{ in } \Gamma\}$ as the case would be (Γ is the boundary of Ω .) We note that in this case $\overline{D(\varphi)} = \{y \in L^2(\Omega); y(x) \in \overline{D(\beta)} \text{ a.e. } x \in \Omega\}$.
Let U be a closed, convex and bounded subset of H containing the origin and let

$$(1.7) \quad \mathcal{U} = \{u \in L^\infty(R^+;H); u(t) \in U \quad \text{a.e. } t > o\}.$$

A control $u \in \mathcal{U}$ is called <u>admissible</u> if it steers y_0 to origin in some time T (if any). The smallest time T for which $y(T,y_0,u) = o$ is called the <u>transition time</u> of u and the infimum $T(y_0)$ of all transition times is called the <u>optimal time</u>, i.e.,

$$T(y_0) = \inf\{T > o;\ y(T,y_0,u) = o,\ u \in \mathcal{U}\}.$$

A control $u \in \mathcal{U}$ for which $y(T(y_0),y_0,u) = o$ (if any) is called <u>time optimal control</u> and the pair $(y(\cdot,y_0,u),u)$ <u>time optimal pair</u>.

It turns out that if the set of admissible controls is nonempty then there exists at least one <u>time optimal control</u> (see for instance [6], Proposition 7.1). This happens in many notable situations and in particular in that described in Lemma 1 below.

LEMMA 1 <u>Let</u> $A = -\Delta$, $D(A) = H_0^1(\Omega) \cap H^2(\Omega)$ <u>and let</u> $U \subset L^\infty(\Omega)$ <u>be such that</u> $o \in$ <u>int</u> U (<u>the interior is taken in</u> $L^\infty(\Omega)$ - <u>topology</u>). <u>Then for every</u> $y_0 \in \overline{D(\varphi)} \cap L^\infty(\Omega)$ <u>there exists at least one admissible control</u>.

In particular it follows under conditions of Lemma 1 that there exists at least one time optimal control.

<u>Proof of Lemma 1</u>. Without loss of generality we may assume that

$$(1.8) \quad U = \{y \in L^\infty(\Omega); |y|_\infty \le r\}$$

where $|\cdot|_\infty$ is the usual L^∞ - norm on Ω .
First we will prove the existence of admissible controls in the case where $U = U_n = \{y \in L^n(\Omega);\ |y|_n^n \le r^n m(\Omega)\}$ where $|\cdot|_n$ is the

$L^n(\Omega)$ - norm and $m(\Omega)$ is the Lebesgue measure of Ω. To this purpose consider the Cauchy problem

$$(1.9)\quad \begin{aligned} &y'+Ay+Fy+r(m(\Omega))^{1/n}\Gamma_n y \ni o \qquad \text{a.e. } t>o \\ &y(o) = y_o \end{aligned}$$

where $\Gamma_n y = y|y|_n^{-1}$ if $y \neq o$ and $\Gamma_n o = \{z \in L^n(\Omega); |z|_n \leq 1\}$. The operator Γ_n is m-accretive in $L^n(\Omega)$ and according to Theorem 3.5 in [2] so is $A+F+r(m(\Omega))^{1/n}\Gamma_n$. (Under our assumptions the operator $A+F$ with the domain $D_n = \{y \in W_o^{1,n}(\Omega) \cap W^{2,n}(\Omega); F(y) \in L^n(\Omega)\}$ is m-accretive in $L^n(\Omega)$.). Then for $y \in D_n$, Eq.(1.9) has a unique strong solution y_n. Multiplying Eq.(1.9) by $y_n|y_n|^{n-2}/|y_n|_n^{n-2}$, integrating on $\Omega \times R^+$ we get the estimate

$$|y_n(t)|_n + r(m(\Omega))^{1/n} t \leq |y_o|_n \quad \forall\ t \geq o.$$

Hence $y_n(t) = o$ for $t \geq T_n = |y_o|_n\, r^{-1}(m(\Omega))^{-\frac{1}{n}}$. We set

$$u_n = -r(m(\Omega))^{1/n}\Gamma_n y_n \text{ for } t \in [o,T_n];\ u_n = o \text{ for } t > T_n$$

and note that u_n is an admissible control for time optimal problem where $U = U_n$. Let us assume that $y_o \in \bigcap_{n=1} D_n$. We have

$$(1.10)\quad (m(\Omega))^{-1/p}\ |u_n|_p \leq (m(\Omega))^{-1/n}\ |u_n|_n \leq r \text{ for } n \geq p.$$

Hence $\{u_n\}$ is weakly compact in every $L^p(\Omega)$ and by (1.10) we see that

$$(1.11)\quad |u|_p \leq r(m(\Omega))^{1/p} \quad \forall\ p \geq 2$$

where u is a weak limit point of $\{u_n\}$ in every $L^p(\Omega)$. By (1.11) we infer that $|u|_\infty \leq r$.

Now letting n tend to $+\infty$ in the equation

$$(1.12)\quad \begin{aligned} &y_n'+Ay_n+Fy_n \ni u_n, \quad o \leq t \leq T, \\ &y_n(o) = y_o \end{aligned}$$

we conclude by a standard device that u is an admissible control with the transition time $|y_o|_\infty r^{-1}$. Clearly the above discussion extend by density to all $y_o \in L^\infty(\Omega) \cap \overline{D(\varphi)}$.

There exists an extensive literature on time optimal problem for linear evolution equations in Banach spaces mainly concerned with the maximum principle (see on these lines [1], [10], [11], [13], [16] and the bibliography given there). The time optimal control problem for the system (1.1) has been studied in [3], [5] (see also Chapter VII in [6]). Here we shall give a general presentation of these results in a more general context with special emphasis on dynamic programming approach.

2. APPROXIMATING TIME OPTIMAL CONTROLS

The idea, previously used in [3], [4], is to approximate the time optimal control problem by the infinite horizon control problem

$$(2.1)\quad \varphi_\varepsilon(y_0)=\inf\left\{\int_0^\infty (g_\varepsilon(y(t))+h_\varepsilon(u(t)))dt;y'+Ay+F^\varepsilon y=u,\ y(0)=y_0\right\}$$

where

$$(2.2)\quad h_\varepsilon(u)=\inf\left\{\frac{|u-v|_2^2}{2\varepsilon}\ ;\ v\in U\right\},\ u\in U$$

$$(2.3)\quad g_\varepsilon(y)=\widetilde{\Pi}(|y|_2^2\ \varepsilon^{-\frac{1}{2}}),\quad y\in H.$$

Herein $\widetilde{\Pi}$ is a C^1- real valued function on R^+ such that $0\le\widetilde{\Pi}\le 1$, $\widetilde{\Pi}(r)=0$ for $0\le r\le 1$, $\widetilde{\Pi}(r)=1$ for $r\ge 2$ and $\widetilde{\Pi}'\ge 0$.

The smooth operator $F^\varepsilon:H\longrightarrow H$ is defined by

$$(2.4)\quad F^\varepsilon(y)(x)=\beta^\varepsilon(y(x))\quad \text{a.e. } x\in\Omega,\ y\in H$$

where $\beta^\varepsilon\in C^1(R)$ is a monotonically increasing Lipschitzian function such that

$$(2.5)\quad |\beta^\varepsilon(r)-\beta_\varepsilon(r)|\le C\varepsilon\quad \forall\,\varepsilon>0,\ r\in R.$$

For instance we may take β^ε as

$$(2.6)\quad \beta^\varepsilon(r)=\int(\beta_\varepsilon(r-\varepsilon^2\theta)\rho(\theta)-\beta_\varepsilon(-\varepsilon^2\theta))\rho(\theta)d\theta$$

where $\rho\in C_0^1(R)$ is such that $\rho(r)=0$ for $|r|>1$ and $\int\rho(r)dr=1$.

LEMMA 2 <u>The function $\varphi_\varepsilon:H\longrightarrow R$ is everywhere finite and locally Lipschitz on H. For every $\varepsilon>0$ and $y_0\in H$ the infimum defining $\varphi_\varepsilon(y_0)$ is attained.</u>

<u>Proof</u>. It is readily seen that $u=0$ is an admissible control in problem (2.1). Indeed by assumption (1.2)

$$|y(t,y_0,0)|_2\le\exp(-\omega t)|y_0|_2\quad \forall\ t\ge 0$$

and therefore $g_\varepsilon(y(t,y_0,0))\in L^1(R^+)$. Hence $\varphi_0(y_0)<+\infty$ for all $y_0\in H$. Now let y_0 be arbitrary but fixed in H. Then there exist the sequences $\{u_n\}$ and $\{y_n\}$ such that

$$(2.7)\quad \varphi_\varepsilon(y_0)\le\int_0^\infty(g_\varepsilon(y_n)+h_\varepsilon(u_n))dt\le\varphi_\varepsilon(y_0)+n^{-1}$$

(2.8) $y_n'+Ay_n+F^{\varepsilon}y_n = u_n$ a.e. $t>0$

$y_n(0) = y_0.$

By (2.2) we see that $\{u_n\}$ remain in a bounded subset of $L^2_{loc}(R^+;H)$. Then by (2.8) it follows that $\{y_n\}$ is bounded in all $C([0,T];H) \cap W^{1,2}([\delta,T];H)\cap L^2(\delta,T;D(A))$; $0<\delta<T<\infty$. Thus selecting a subsequence, if necessary, we may assume that

$u_n \longrightarrow u_{\varepsilon}$ weakly in every $L^2(0,T;H)$

$y_n \longrightarrow y_{\varepsilon}$ strongly in every $L^2(0,T;H)$
and weakly in all $W^{1,2}([\delta,T];H)$.

The latter implies via Fatou's lemma

(2.9) $\varphi_{\varepsilon}(y_0) = \int_0^{\infty}(g_{\varepsilon}(y_{\varepsilon})+h_{\varepsilon}(u_{\varepsilon}))dt$

as claimed.

Finally, if y_0 and z_0 are arbitrary but fixed in H we have

$$\varphi_{\varepsilon}(z_0)-\varphi_{\varepsilon}(y_0)\leq \int_0^{\infty}(g_{\varepsilon}(\tilde{y}_{\varepsilon}(t))-g_{\varepsilon}(y_{\varepsilon}(t)))dt$$

where

$\tilde{y}_{\varepsilon}'+A\tilde{y}_{\varepsilon}+F^{\varepsilon}\tilde{y}_{\varepsilon} = u_{\varepsilon}$ a.e. in R^+

$\tilde{y}_{\varepsilon}(0) = z_0$

and $(y_{\varepsilon},u_{\varepsilon})$ satisfy (2.9). Recalling that

$$|y_{\varepsilon}(t)-\tilde{y}_{\varepsilon}(t)|_2 \leq \exp(-\omega t)|y_0-z_0|_2 \quad \forall\ t\geq 0$$

we find that

$$\varphi_{\varepsilon}(z_0)-\varphi_{\varepsilon}(y_0)\leq L_{\varepsilon}|y_0-z_0|_2$$

thereby completing the proof.

LEMMA 3. _Assume that_ $y_0\in D(\varphi)$ _and that the time optimal control problem admits at least one admissible control_. _Then there exists a sequence_ $\varepsilon \longrightarrow 0$ _such that_

(2.10) $\varphi_{\varepsilon}(y_0) \longrightarrow T^* = T(y_0)$

(2.11) $u_{\varepsilon} \longrightarrow u^*$ weak star in $L^{\infty}(0,T^*;H)$

(2.12) $y_{\varepsilon} \longrightarrow y^*$ weakly in $W^{1,2}([0,T^*];H)\cap L^2(0,T^*;D(A))$
and strongly in $C([0,T^*];H)$

(2.13) $F^{\varepsilon}(y_{\varepsilon}) \longrightarrow \xi \in Fy^*$ weakly in $L^2(0,T^*;H)$.

Here (u^*,y^*) _is a time optimal pair and_ $(u_{\varepsilon},y_{\varepsilon})$ _is an optimal pair in problem_ (2.1).

<u>Proof</u>. Since the proof is identic with that of Lemma 7.3 in [6] (see also [5]) it will be outlined only. By assumptions there exists at least one time optimal pair (y_1^*, u_1^*). We extend u_1^* and y_1^* by o outside interval $[o,T^*]$ and by uniqueness of the Cauchy problem we see that

(2.14) $y_1^{*'} + Ay_1^* + Fy_1^* \ni u_1^*$ a.e. $t > o$.

On the other hand, we have

(2.15) $\varphi_\varepsilon(y_0) = \int_0^\infty (g_\varepsilon(y_\varepsilon) + h_\varepsilon(u_\varepsilon))dt \le \int_0^\infty g_\varepsilon(z_\varepsilon)dt$

where

(2.16) $z_\varepsilon' + Az_\varepsilon + F^\varepsilon z_\varepsilon = u_1^*$ a.e. in R^+

$z_\varepsilon(o) = y_0$.

Then again using condition (1.2) we get the estimate

$$|z_\varepsilon(t)|_2 \le \exp(-\omega(t-T^*))|z_\varepsilon(T^*)|_2 \quad \forall\ t \geqslant T^*$$

whilst by (2.5) and (2.14) we have

$$|z_\varepsilon(t) - y_1^*(t)|_2 \le C\varepsilon^{1/2} \quad \forall\ t \in [o,T^*].$$

Hence

$$|z_\varepsilon(t)|_2 \le C\ \varepsilon^{1/2} \quad \text{for all} \quad t \geqslant T^*$$

and by (2.15) we see that for all $\varepsilon > o$ and sufficiently small

(2.17) $\varphi_\varepsilon(y_0) \le T^*$.

On the other hand, inasmuch as in virtue of (2.2) and (2.15), $\{u_\varepsilon\}$ is bounded in $L^2(o,T^*;H)$, there exists a sequence convergent to zero, again denoted ε, such that

$u_\varepsilon \longrightarrow u^*$ weakly in $L^2(o,T;H)$

$y_\varepsilon \longrightarrow y^*$ weakly in $W^{1,2}([o,T];H) \cap L^2(o,T;D(A))$ and strongly in $C([o,T];H)$

for every $T > o$, where $y^* = y(\cdot, y_0, u^*)$.

We will show that u^* is a time optimal control To this purpose note first by (2.17) that $m\{t \geqslant o; |y_\varepsilon(t)|_2^2 \geqslant 2\ \varepsilon^{1/2}\} \le T^*$ (m is the Lebesgue measure). Thus there exists $\{\varepsilon_n\} \longrightarrow o$ and $\{t_n\} \subset [o,2T^*]$ such that

$$|y_{\varepsilon_n}(t_n)|_2^2 \le 2\ \varepsilon_n^{1/2} \quad \text{for all } n.$$

Without loss of generality we shall assume that $t_n \longrightarrow T_0$. Since

$\{y'_{\varepsilon_n}\}$ is bounded in $L^2(o,2T^*;H)$ we have

$$|y_{\varepsilon_n}(t)-y_{\varepsilon_n}(t_n)|_2 \leq C|t-t_n|^{1/2} \quad \forall\, t\in[o,2T^*].$$

Hence $y^*(T_o) = o$. Now let $\tilde{T}=\inf\{T; y^*(T)=o\}$. We will prove that $\tilde{T}=T^*$. To this end we set

$$E_\varepsilon = \{t\in[o,\tilde{T}]; |y_\varepsilon(t)|_2^2 \geq 2\,\varepsilon^{1/2}\}.$$

By (2.17) it follows that $m(E_\varepsilon)\leq T^*\leq \tilde{T}$. Let us prove that $\lim_{\varepsilon\downarrow o} m(E_\varepsilon) \leq \tilde{T}$. Indeed otherwise there would exist $\delta > o$ and $\varepsilon_n \to o$ such that $m(E_{\varepsilon_n})\leq \tilde{T}-\delta$. In other words, there would exist a sequence $\{A_n\}$ of measurable subsets of $[o,\tilde{T}]$ such that $m(A_n)\geq\delta$ and

$$|y_{\varepsilon_n}(t)|_2^2 \leq 2\,\varepsilon_n^{1/2} \quad \forall\, t\in A_n.$$

Hence

$$|y^*(t)|_2^2 \leq 2\,\varepsilon_n^{1/2} + \nu_n \quad \text{for } t\in A_n$$

where $\nu_n \to o$ for $n\to\infty$. Since $y^*(t)\neq o$ for $t\in[o,\tilde{T}[$ we arrived at a contradiction. We have therefore proved that $\lim_{\varepsilon\downarrow o} m(E_\varepsilon) = T^* = \tilde{T}$ and so u^* is a time optimal control. Since by (2.3) and (2.17), $m(E_\varepsilon)\leq \varphi_\varepsilon(y_o)\leq T^*$, (2.10) follows and the proof is complete.

REMARK 1 If $y_o\in\overline{D(\varphi)}$ then Lemma 3 still remains valid except that (2.12) becomes

(2.12)' $\quad y_\varepsilon \longrightarrow y^*$ weakly in every $W^{1,2}([\delta,T^*];H)\cap L^2(\delta,T^*;D(A))$ and strongly in $C(]o,T^*];H)\cap L^2(o,T^*;H)$.

By (2.1) and (2.2) we see that $u_\varepsilon\in L^\infty(R^+;H)+L^2(R^+;H)$. Then $G_\varepsilon(y_\varepsilon)=2y_\varepsilon\,\varepsilon^{-1/2}\,\pi'(|y_\varepsilon|_2^2\,\varepsilon^{-1/2})\in L^\infty(R^+;H)$ and so there is a unique function $p_\varepsilon\in C(R^+;H)\cap L^\infty(R^+;H)\cap W^{1,2}_{loc}(R^+;H)\cap L^2_{loc}(R^+;D(A))$ which satisfies the equation

(2.18) $\quad p'_\varepsilon - Ap_\varepsilon - p_\varepsilon\,\dot{\beta}^\varepsilon(y_\varepsilon)=G_\varepsilon(y_\varepsilon) \quad$ a.e. $t>o$.

Finally, since g_ε and h_ε are Fréchet differentiable it follows by a standard device that

(2.19) $\quad p_\varepsilon = \nabla h_\varepsilon(u_\varepsilon) \quad$ a.e. $t>o$

or equivalently

(2.20) $\quad u_\varepsilon\in\partial h^*_\varepsilon(p_\varepsilon) \quad$ a.e. $t>o$

where h^*_ε is the conjugate of h_ε.

To summarize, we have shown that <u>there exist the sequences</u> $\{y_\varepsilon\}$, $\{p_\varepsilon\}$, $\{u_\varepsilon\}$ <u>satisfying</u> (2.11), (2.12), (2.13) <u>and</u> Eqs.(2.18), (2.19) <u>and</u> (2.21) <u>below</u>

$$(2.21)\quad \begin{aligned} & y_\varepsilon' + Ay_\varepsilon + F^\varepsilon y_\varepsilon = u_\varepsilon \quad \text{a.e. } t>0 \\ & y_\varepsilon(0) = y_0. \end{aligned}$$

Eqs.(2.21), (2.18) and (2.19) taken together represent first order conditions of optimality for problem (2.1).

3. THE DYNAMIC PROGRAMMING EQUATION AND FEEDBACK CONTROLS.

We shall prove here the following theorem

THEOREM 1. <u>The optimal value function</u> φ_ε <u>defined by</u> (2.1) <u>satisfies the equation</u>

$$(3.1)\quad (Ay+F^\varepsilon y, \partial\varphi_\varepsilon(y)) + h_\varepsilon^*(-\partial\varphi_\varepsilon(y)) = g_\varepsilon(y) \quad \forall\, y\in D(A)$$

<u>and every optimal control</u> u_ε <u>of problem</u> (2.1) <u>is given as a function of optimal state</u> y_ε <u>by the feedback law</u>

$$(3.2)\quad u_\varepsilon(t) \in \partial h_\varepsilon^*(-\partial\varphi_\varepsilon(y_\varepsilon(t))) \quad \text{a.e. } t>0.$$

Herein we have denoted by $\partial\varphi_\varepsilon : H \longrightarrow 2^H$ the generalized gradient of φ_ε ([12]), i.e.,

$$\partial\varphi_\varepsilon(y) = \{w\in H; (w,h) \le \varphi_\varepsilon^0(y,h) \quad \forall\, h\in H\}$$

$$\varphi_\varepsilon^0(y) = \limsup_{\substack{z\to y \\ \lambda\downarrow 0}} \lambda^{-1}(\varphi_\varepsilon(z+\lambda h) - \varphi_\varepsilon(z)).$$

Eq.(3.1) should be understood in the following sense: for each $y\in D(A)$ there is $\eta_\varepsilon\in\partial\varphi_\varepsilon(y)$ such that

$$(3.3)\quad (Ay+F^\varepsilon y, \eta_\varepsilon) + h_\varepsilon^*(-\eta_\varepsilon) = g_\varepsilon(y).$$

<u>Proof of Theorem 1</u>. Multiplying Eq.(2.21) by p_ε', Eq.(2.18) by y_ε' and subtracting the results we get

$$(3.4)\quad \frac{d}{dt}((Ay_\varepsilon(t)+F^\varepsilon y_\varepsilon(t), p_\varepsilon(t)) - h_\varepsilon^*(p_\varepsilon(t)) + g_\varepsilon(y_\varepsilon(t))) = 0 \text{ a.e. } t>0.$$

Since by (2.19), $p_\varepsilon \in L^2(R^+;H)$ and h_ε^* is continuous we infer by the obvious inequality

$$\int_0^t |Ay_\varepsilon(s)+F^\varepsilon y_\varepsilon(s)|^2 ds \le C(t+1) \quad \forall\, t\ge 0$$

that (see [3], [5], [6])

(3.5) $\lim_{t_n \to \infty} (Ay_\varepsilon(t_n)+F^\varepsilon y_\varepsilon(t_n),p_\varepsilon(t_n)) = 0$

for some $t_n \to +\infty$. On the other hand, since $g_\varepsilon(y_\varepsilon) \in L^1(R^+)$ and $t \to g_\varepsilon(y_\varepsilon(t))$ is uniformly continuous we conclude that $\lim_{t\to\infty} g_\varepsilon(y_\varepsilon(t)) = 0$. The latter combined with (3.4) and (3.5) yields

(3.6) $(Ay_\varepsilon(t)+F^\varepsilon y_\varepsilon(t),p_\varepsilon(t))-h_\varepsilon^*(p_\varepsilon(t))+g_\varepsilon(y_\varepsilon(t))=0 \quad \forall\ t>0.$

(We note that the function $t \to (Ay_\varepsilon(t)+F^\varepsilon y_\varepsilon(t),p_\varepsilon(t))$ is absolutely continuous and in particular continuous on $]0,+\infty[$.) If $y_\varepsilon(0) = y_0 \in D(A)$ then (3.6) yields

(3.7) $(Ay_0+F^\varepsilon y_0,p_\varepsilon(0))-h_\varepsilon^*(p_\varepsilon(0)) = g_\varepsilon(y_0)$.

On the other hand, it is readily seen that

$$\varphi_\varepsilon(y_0) = \inf\Big\{ \int_0^t (g_\varepsilon(y(s))+h_\varepsilon(u(s)))ds+\varphi_\varepsilon(y(t));$$
$$y'+Ay+F^\varepsilon y=u \quad \text{a.e. } s\in[0,t],\ y(0)=y_0\Big\}.$$

Then by the maximum principle (see [3] or [6] Chap.5) for every $t>0$ there exists $p_\varepsilon^t \in W^{1,2}([0,t];H)\cap L^2(0,t;D(A))$ such that

(3.8) $(p_\varepsilon^t)' - Ap_\varepsilon^t - p_\varepsilon^t \beta^\varepsilon(y_\varepsilon) = G_\varepsilon(y_\varepsilon) \quad$ a.e. in $[0,t]$

(3.9) $p_\varepsilon^t(t) \in -\partial\varphi_\varepsilon(y_\varepsilon(t))$

(3.10) $p_\varepsilon^t(s) = \partial h_\varepsilon^*(u_\varepsilon(s)) \quad \forall\ s\in[0,t].$

Here u_ε is an arbitrary optimal arc of problem (2.1). Comparing Eqs.(2.19) and (3.10) we see that $p_\varepsilon = p_\varepsilon^t$ in $[0,t]$ and by (3.9) we have

(3.11) $p_\varepsilon(t) \in -\partial\varphi_\varepsilon(y_\varepsilon(t)) \quad \forall\ t\geq 0.$

In particular, it follows that $p_\varepsilon(0) \in -\partial\varphi_\varepsilon(y_0)$. Substituting the latter in Eq.(3.7) we obtain (3.1) or (3.3) as claimed. Finally, by (2.20) and (3.11) follows (3.2) and the proof is complete.
Keeping in mind that by Lemma 2, $\varphi_\varepsilon \to T$ letting ε tend to zero (formally) in Eq.(3.1) we may regard $T = T(y)$ as generalized solution to the Bellman equation

(3.12) $(Ay+Fy,\partial T(y))+h^*(-\partial T(y))=1 \quad \forall\ y \neq 0$

where $h^*(p) = \sup\{(p,v);v\in U\}$ is the support function of U.

Now we shall discuss the existence of optimal feedback controls for the time optimal control problem. We restrict ourselves to the case where β is locally Lipschitz and satisfies the growth condition

(3.13) $o \leq \beta'(r) \leq C(|\beta(r)|+|r|+1)$ a.e. $r \in R$

and U is given by

(3.14) $U = U_2 = \{y \in H; |y|_2 \leq 1\}$.

Then the optimal time function T is everywhere defined and locally Lipschitz on H([10]). Hence the generalized gradient $\partial T(y)$ is well defined for every $y \in H$. It turns out that in the present situation every time optimal can be expressed in feedback form. More precisely we have

THEOREM 2. Let u^* be any time optimal control for the system (1.1) with control constraint set (3.14). Then we have

(3.15) $u^*(t) \in -\underline{sgn}\,\partial T(y^*(t))$ a.e. $t \in [o,T^*]$

where y^* is the corresponding optimal state and

$\underline{sgn}\ p = p|p|_2^{-1}$ if $p \neq o$; $\underline{sgn}\ o = \{w \in H; |w|_2 \leq 1\}$.

Proof. Let (u^*,y^*) be any time optimal pair. It is readily seen that for every $o < t \leq T^*$, (u^*,y^*) is an optimal pair for the optimal control problem

$$T(y_o) = \inf\left\{\int_o^t ds + T(y(t)); y' + Ay + Fy \ni u \quad \text{in } [o,t],\ y(o) = y_o\right\}.$$

Then by the maximum principle, there is $p^t \in W^{1,2}([o,t];H) \cap L^2(o,t;D(A))$ such that

(3.16) $(p^t)' - Ap^t - p^t\partial\beta(y^*) \ni o$ a.e. in $[o,t]$

(3.17) $p^t(t) \in -\partial T(y^*(t))$

(3.18) $u^*(s) \in \underline{sgn}\ p^t(s)$ a.e. $s \in [o,t]$.

Arguing as in the proof of Theorem 1 in [3] (see also Theorem 5.6 in [6], Chap.5) we infer that

$$u^*(t) \in \underline{sgn}\ p^t(t) \quad \text{a.e. } t \in [o,T^*]$$

which along with (3.17) yields (3.15) as claimed.

REMARK 2 Theorem 2 remains valid (under controllability assumptions) for more general systems of the form

$$y' + Ay + Fy = Bu$$

where B is a linear continuous operator from a control space U to H. It should be observed by Eq.(3.15) that T satisfies in certain weak sense the dynamic programming equation (3.12), i.e.,

(3.19) $(Ay+Fy, \partial T(y)) + |\partial T(y)|_2 = 1 \quad \forall\, y \neq 0.$

Related Hamilton-Jacobi equations have been studied in [18] (see also [8]) but the treatement does not cover the present case. Since the numerical treatement of this equation seems to be impossible at this stage it would be desirable to obtain suboptimal feedback laws starting from Eq.(3.1).

4. THE MAXIMUM PRINCIPLE

Here we shall derive first order necessary conditions of optimality for the time optimal control problem associated with system (1.1) in the case where A is a second order linear elliptic operator on Ω and U has nonempty interior in $L^\infty(\Omega)$. To be more specific we shall assume that

$$A = -\Delta,\ D(A) = H_0^1(\Omega) \cap H^2(\Omega)$$

$$U = \{u \in L^2(\Omega);\, |u(x)| \leq 1 \quad \text{a.e. } x \in \Omega\}.$$

As seen in Introduction (Lemma 1) for every $y_0 \in L^\infty(\Omega) \cap \overline{D(\varphi)}$ the time optimal control problem admits in this case at least one optimal control. Throughout this section we shall assume that

(4.1) $y_0 \in D(\varphi) \cap W_0^{2-\frac{2}{q},q}(\Omega) \cap L^\infty(\Omega) \quad$ where $q > N$.

Now we return back to the sequences $\{y_\varepsilon\}$, $\{u_\varepsilon\}$, $\{p_\varepsilon\}$ found in Section 2 and note that by Eq.(2.18) we have

(4.2)
$$\frac{\partial p_\varepsilon}{\partial t} + \Delta p_\varepsilon - p_\varepsilon \dot{\beta}^\varepsilon(y_\varepsilon) = 0 \text{ in } Q_\lambda = \Omega \times]0, T_\lambda^*[$$
$$p_\varepsilon = 0 \qquad \text{in } \Sigma_\lambda = \Gamma \times]0, T_\lambda^*[$$

for every $\lambda > 0$ and $0 < \varepsilon < \varepsilon_0(\lambda)$. (Here $T_\lambda^* = T^* - \lambda$.)

Now by a little calculation involving Eqs.(2.18), (2.19) and (2.21) we get

(4.3) $$\frac{d}{dt}(p_\varepsilon(t), Ay_\varepsilon(t) + F^\varepsilon y_\varepsilon(t)) = (Ay_\varepsilon(t) + F^\varepsilon y_\varepsilon(t), G_\varepsilon(y_\varepsilon(t))) +$$
$$+ (\partial h_\varepsilon^*(p_\varepsilon(t)), Ap_\varepsilon(t) + p_\varepsilon(t)\dot{\beta}^\varepsilon(y_\varepsilon(t))) \geq 0 \quad \text{a.e. } t > 0$$

because

(4.4) $h_\varepsilon^*(p) = |p|_1 + \frac{\varepsilon}{2}|p|_2^2 \quad \forall\ p \in L^2(\Omega)$

and

(4.5) $\partial h_\varepsilon^*(p)(x) = \text{sgn}\ p(x) + \varepsilon p(x) \quad$ a.e. $x \in \Omega$.

Here $|\cdot|_1$ is the L^1-norm and $\text{sgn}\ p = p|p|^{-1}$ if $p \neq o$, $\text{sgn}\ o = [-1,1]$.

Along with (3.5), (3.6) and (3.4), (4.3) yields

(4.6) $|p_\varepsilon(t)|_1 + \frac{\varepsilon}{2}|p_\varepsilon(t)|_2^2 \leq g_\varepsilon(y_\varepsilon(t)) \leq 1 \quad \forall\ t \geq o.$

Now multiply Eq.(4.2) by $\text{sgn}\ p_\varepsilon$ and integrate on $\Omega \times [o,t]$ to get

(4.7) $|p_\varepsilon(t)|_1 + \int_o^t \int_\Omega |\dot\beta^\varepsilon(y_\varepsilon)p_\varepsilon|\,dxds \leq 1 \quad \forall\ t \in [o,T_\lambda^*].$

For all $h_i \in L^2(o,T_\lambda^*;L^q(\Omega))$, $i = 1,2,\dots N$ the boundary value problem

(4.8) $\frac{\partial v}{\partial t} - \Delta v = \sum_{i=1}^N (h_i)_{x_i} \quad$ in Q_λ

$v(x,o) = o$ for $x \in \Omega$; $v = o$ in Σ_λ,

has a unique solution $v \in L^2(o,T_\lambda^*;H_o^1(\Omega))$ with $v_t \in L^2(o,T_\lambda^*;H^{-1}(\Omega))$. Moreover, if $q > N$ then $v \in L^\infty(Q_\lambda)$ and (see [14], Chap.III, Theorem 7.1)

(4.9) $\|v\|_{L^\infty(Q_\lambda)} \leq C \sum_{i=1}^N \|h_i\|_{L^2(o,T_\lambda^*;L^q(\Omega))}$.

Now multiplying Eq.(4.8) by $|v|^{p-2}v$ and integrating on $\Omega \times]o,t[$ we find after some manipulations that

(4.10) $|v(\cdot,t)|_\infty \leq C\|v\|_{L^\infty(Q_\lambda)}$.

Then multiplying Eq.(4.2) by v and integrating on Q_λ we find after some calculation that

$$\left|\sum_{i=1}^N \int_{Q_\lambda} (p_\varepsilon)_{x_i} h_i\,dxdt\right| \leq C \sum_{i=1}^N \|h_i\|_{L^2(o,T_\lambda^*;L^q(\Omega))}.$$

(We denote by C several positive constants independent of ε and λ.)

Hence

(4.11) $\|p_\varepsilon\|_{L^2(o,T_\lambda^*;W_o^{1,q'}(\Omega))} \leq C \quad \forall\ \varepsilon, \lambda > o,$

where $q^{-1}+(q')^{-1}=1$. In particular, it follows that $\{p_\varepsilon\}$ is bounded in $L^2(o,T^*_\lambda;L^{q'}(\Omega))$ and $\{(p_\varepsilon)_t\}$ is bounded in $L^1(o,T^*_\lambda;H^{-s}(\Omega)+W^{-1,q}(\Omega))$ where $s>N/2$. We set $Y^*=H^{-s}(\Omega)+W^{-1,q}(\Omega)$ and note that in virtue of estimate (4.7), $\{p_\varepsilon(t);\ o<\varepsilon<\varepsilon_o(\lambda)\}$ is for every $t\in[o,T^*_\lambda]$ a precompact subset of Y^*. Then according to the vectorial version of the Helly theorem, there exists a function $p\in BV([o,T^*[;Y^*)\cap L^2(o,T^*;W_o^{1,q'}(\Omega))$ and a subsequence $\varepsilon_n\to o$ such that

(4.12) $p_{\varepsilon_n}(t)\longrightarrow p(t)$ strongly in Y^* $\forall t\in[o,T^*[$,

(4.13) $p_{\varepsilon_n}\longrightarrow p$ weakly in $L^2(o,T^*;W_o^{1,q'}(\Omega))$.

Arguing as in the proof of Proposition 7.2 in [6] we may extend p as a function of bounded variation on the closed interval $[o,T^*]$. (We have denoted by $BV([o,T^*[;Y^*)$ the space of Y^*-valued functions of bounded variation on $[o,T^*[$.)

On the other hand, according to Lemma 3.1 in [15], Chap.I, for every $\eta>o$ there exists $C(\eta)>o$ such that

$$|p_{\varepsilon_n}(t)-p(t)|_{q'}\leq\eta\|p_{\varepsilon_n}(t)-p(t)\|_{W_o^{1,q'}(\Omega)}+$$

$$+C(\eta)\|p_{\varepsilon_n}(t)-p(t)\|_{Y^*}\quad\forall\ t\in[o,T^*_\lambda].$$

Hence

(4.14) $p_{\varepsilon_n}\longrightarrow p$ strongly in $L^2(o,T^*;L^{q'}(\Omega))$.

Now by Eq.(3.7) we have for $o<\varepsilon<\varepsilon_o(\lambda)$

(4.15) $|p_\varepsilon(t)|_1+\frac{\varepsilon}{2}|p_\varepsilon(t)|_2^2=(p_\varepsilon(t),Ay_\varepsilon(t)+F^\varepsilon y_\varepsilon(t))+1$ a.e. $t\in[o,T^*_\lambda]$.

Keeping in mind that

$$h_\varepsilon(u_\varepsilon)+h^*_\varepsilon(p_\varepsilon)=(u_\varepsilon,p_\varepsilon)=|p_\varepsilon|_1+\varepsilon|p_\varepsilon|_2^2$$

and noting that by Lemma 3

$$\int_o^\infty h_\varepsilon(u_\varepsilon)dt\longrightarrow o\qquad\text{for}\quad\varepsilon\to o$$

we infer that

$$\varepsilon|p_\varepsilon(t)|_2^2\longrightarrow o\ \text{ strongly in } L^1(R^+)\ \text{for}\ \ \varepsilon\to o.$$

Then (4.15) yields

(4.16) $|p(t)|_1 - \lim_{\varepsilon \to 0} (p_\varepsilon(t), Ay_\varepsilon(t)+F^\varepsilon y_\varepsilon(t)) = 1$, a.e. $t \in [0,T^*]$.

Multiplying Eq.(2.21) by $|\beta^\varepsilon(y_\varepsilon)|^{q-2}\beta^\varepsilon(y_\varepsilon)$ and integrating on Q we find that $\{F^\varepsilon y_\varepsilon\}$ is bounded in $L^q(Q)$ and using Theorem 9.1 in [14] we conclude that

(4.17) $\|y_\varepsilon\|_{W_q^{2,1}(Q)} \leq C \quad \forall \varepsilon > 0.$

Then we may pass to limit in Eq.(4.16) to get

(4.18) $|p(t)|_1 - (Ay(t)+Fy(t), p(t)) = 1$, a.e. $t \in [0,T^*]$.

Next it follows by estimate (4.7) that there is $\mu \in (L^\infty(Q))^*$ and a generalized subsequence of $\{\dot\beta^\varepsilon(y_\varepsilon)p_\varepsilon\}$ such that

(4.19) $\dot\beta^\nu(y_\nu)p_\nu \longrightarrow \mu$ weak star in $(L^\infty(Q_\lambda))^*$.

Now by Eq.(2.20), we have

(4.20) $u_\varepsilon = \varepsilon p_\varepsilon + \mathrm{sgn}\, p_\varepsilon$ a.e. in Q.

To summarize, we have shown that <u>there exists</u> $p \in L^2(0,T; W_0^{1,q'}(\Omega)) \cap BV([0,T^*];Y^*)$ <u>and</u> $\mu \in (L^\infty(Q))^*$ <u>such that</u>

(4.21) $p_t + \Delta p = \mu$ in Q ,

(4.22) $u^* \in \mathrm{sgn}\, p$ a.e. in Q .

(The latter follows letting ε tend to zero in Eq.(4.20).)

In the special case where $\mu = 0$ (i.e., $\beta \equiv 0$) it is known that the set of zeros of p in Q is of Lebesgue measure zero and so u^* is a bang-bang control. In this case Eqs.(4.18), (4.21), (4.22) represent the maximum principle for the linear heat equation. In the general case we consider here Eqs.(4.18), (4.21), (4.22) together with (4.18) represent a weak form of the maximum principle and it implies in particular that u^* is a bang-bang control on a subset of positive measure.

Eq.(4.21) can be made more explicit in two specific cases we will consider separately.

1° β is locally Lipschitz and satisfies condition (3.13) i.e.,

(4.23) $0 \leq \beta'(r) \leq C(|\beta(r)| + |r| + 1)$.

2° $\beta \subset R \times R$ is the multivalued graph

(4.24) $\beta(r) = 0$ for $r > 0$; $\beta(0) =]-\infty, 0]$, $\beta(r) = \emptyset$ for $r < 0$.

In the case 1°, using the obvious inequality

$$\dot{\beta}^{\varepsilon}(r) \leq C(|\beta^{\varepsilon}(r)| + |r| + 1) \quad \forall\ r \in R$$

we get

$$\int_E |p_\varepsilon \dot{\beta}^{\varepsilon}(y_\varepsilon)|\,dxdt \leq C(\int_E |p_\varepsilon||\beta^{\varepsilon}(y_\varepsilon)|\,dxdt + \int_E |p_\varepsilon||y_\varepsilon|\,dxdt + \int_E |p_\varepsilon|\,dxdt)$$

for every measurable subset E of Q_λ. Since as seen earlier $\{p_{\varepsilon_n}\}$ is strongly convergent in $L^2(o,T^*;L^{q'}(\Omega))$ and $\{\beta^{\varepsilon_n}(y_{\varepsilon_n})\}$ is weakly compact in $L^2(o,T;L^q(\Omega))$, the latter implies via Dunford-Pettis criterion that $\{p_\varepsilon \dot{\beta}^{\varepsilon}(y_\varepsilon)\}$ is a weakly compact subset of $L^1(Q)$. Hence $\mu \in L^1(Q)$ and p is absolutely continuous Y^*-valued function on $[o,T^*]$, i.e., $p \in AC([o,T^*];Y^*)$. On the other hand, by Lemma 3.4 in [6] we infer that

$$\mu(x,t) \in p(x,t)\,\partial\beta(y^*(x,t)), \quad \text{a.e. } (x,t) \in Q$$

where $\partial\beta$ is the generalized gradient of β.

Therefore, we have proved the following theorem

THEOREM 3. <u>Assume that</u> $y_0 \in L^\infty(\Omega) \cap W_0^{2-\frac{2}{q},q}(\Omega)$ <u>and satisfies condition</u> (4.23). <u>Then there exists at least one time optimal pair</u> (y^*,u^*) <u>which satisfies the maximum principle in the following sense: there exists</u> $p \in L^2(o,T^*;W_0^{1,q'}(\Omega)) \cap AC([o,T^*];Y^*)$ <u>with</u> $p_t + \Delta p \in L^1(Q)$, <u>which satisfy the equations</u>

(4.25) $p_t + \Delta p - p\,\partial\beta(y^*) \ni o \quad \text{in } Q = \Omega \times]o,T^*[$

(4.26) $u^* \in \operatorname{sgn} p, \quad \text{a.e. in } Q$

(4.27) $|p(t)|_1 - (Ay(t) + Fy(t), p(t)) = 1, \quad \text{a.e. } t \in [o,T^*].$

If β is defined by (4.24) then Eq.(1.6) reduces to the free boundary value problem

$$y_t - \Delta y = u \quad \text{in } \{(x,t); y(x,t) > o\}$$

(4.28) $y_t - \Delta y \geqslant u, \quad y \geqslant o \text{ in } \Omega \times R^+$

$$y(x,o) = y_0(x) \quad \text{for } x \in \Omega\ ;\ y = o \text{ in } \Gamma \times R^+.$$

To obtain an explicit form of the optimality system we choose β^ε of the form (2.6), i.e.,

$$\beta^{\varepsilon}(r) = \varepsilon^{-1} \int_{r\varepsilon^{-2}}^{1} (r - \varepsilon^2\theta)\rho(\theta)d\theta + \varepsilon\int_0^1 \theta\rho(\theta)d\theta, \quad r \in R.$$

This yields

$$(4.29)\quad p_\varepsilon(\beta^\varepsilon(y_\varepsilon)-y_\varepsilon\dot\beta^\varepsilon(y_\varepsilon)) = -\varepsilon p_\varepsilon\Big(\int_{\varepsilon^{-2}y_\varepsilon}^{1}\theta\rho(\theta)d\theta+\int_{0}^{1}\theta\rho(\theta)d\theta\Big)$$

$$\longrightarrow o \text{ strongly in } L^2(o,T^*;H) \text{ for } \varepsilon\longrightarrow o.$$

On the other hand, we have

$$(4.30)\quad |p_\varepsilon\beta^\varepsilon(y_\varepsilon)|\le|p_\varepsilon\dot\beta^\varepsilon(y_\varepsilon)|\,(\varepsilon^2+\eta_\varepsilon|y_\varepsilon|)+2\varepsilon|p_\varepsilon|$$

where

$$\eta_\varepsilon=\begin{cases}o & \text{if } y_\varepsilon>-\varepsilon^2\\ 1 & \text{if } y_\varepsilon\le-\varepsilon^2.\end{cases}$$

Since by Lemma 3, $\{\varepsilon^{-1}y_\varepsilon\eta_\varepsilon\}$ is bounded in $L^2(Q)$ it follows by (4.30) that on a subsequence, again denoted $\varepsilon_n\longrightarrow o$ we have

$$p_{\varepsilon_n}\beta^{\varepsilon_n}(y_{\varepsilon_n})\longrightarrow o \quad \text{a.e. in } Q=\Omega\times]o,T^*[.$$

Since $\{\beta^\varepsilon(y_\varepsilon)\}$ is weakly compact in $L^2(o,T;L^q(\Omega))$ it follows by (4.13) and (4.14) that

$$p_{\varepsilon_n}\beta^{\varepsilon_n}(y_{\varepsilon_n})\longrightarrow p\upsilon \qquad \text{weakly in } L^1(Q)$$

where $\upsilon=u^*-y_t^*-Ay^*\in\beta(y^*)$ a.e. in Q. Hence $p\upsilon=o$ a.e. in Q and

$$(4.30)\quad p_{\varepsilon_n}\beta^{\varepsilon_n}(y_{\varepsilon_n})\longrightarrow o \quad \text{strongly in } L^1(Q)$$

$$p(u^*-y_t^*-Ay^*)=o \qquad \text{a.e. in } Q.$$

Recalling that

$$u^*\in \operatorname{sgn} p \qquad \text{a.e. in } Q$$

the latter yields

$$(4.31)\quad p(x,t)=o \qquad \text{a.e. in } \{(x,t)\in Q;y^*(x,t)=o\}.$$

Finally, by (4.29) we see that

$$(4.32)\quad p_{\varepsilon_n}y_{\varepsilon_n}\dot\beta^{\varepsilon_n}(y_{\varepsilon_n})\longrightarrow o \text{ strongly in } L^1(Q).$$

On the other hand, by the Egorov theorem, for every $\tau>o$ there exists $E_\tau\subset Q$ such that $m(Q\setminus E_\tau)\le\tau$, y_{ε_n} are uniformly bounded in E_τ and

$$y_{\varepsilon_n}\longrightarrow y^* \quad \text{uniformly in } E_\tau.$$

Along with (4.19) this yields $y^*\mu = o$ in E_τ. In other words,

$$\int_{E_\tau} \mu_a\, y^*\psi\, dxdt + \mu_s(y^*\psi) = o$$

for all $\psi \in L^\infty(Q)$ which vanish outside E_τ.

Here μ_a and μ_s are the absolutely continuous part and the singular part, respectively, of μ. Hence $y^*\mu_a = o$ a.e. in Q. To summarize, we have proved the following theorem

THEOREM 4 <u>Assume that</u> $y_o \in L^\infty(\Omega) \cap W^{2-\frac{2}{q},q}(\Omega)$ <u>and</u> $y_0 \geq o$ a.e. <u>in</u> Ω. <u>Then there exists at least one time optimal control</u> u^* <u>for the system</u> (4.28) <u>which satisfies the maximum principle in the following sense: there exists</u> $p \in BV([o,T^*];Y^*) \cap L^2(o,T^*;L^{q'}(\Omega))$ <u>and</u> $\mu \in (L^\infty(Q))^*$ <u>such that</u>

(4.32) $\quad p_t + \Delta p = \mu \quad$ in Q.

(4.33) $\quad \mu_a = o \quad$ a.e. in $\{(x,t) \in Q; y^*(x,t) > o\}$.

(4.34) $\quad p = o \quad$ a.e. in $\{(x,t) \in Q; y^*(x,t) = o\}$.

(4.35) $\quad u^* \in \mathrm{sgn}\, p$ a.e. in Q.

(4.36) $\quad |p(t)|_1 - (Ay(t)+Fy(t),p(t)) = 1$, a.e. $t \in [o,T^*]$.

Eq.(4.32) is considered in the sense of distributions on Q and taken together with (4.33), (4.34), (4.35) and (4.28) represent a quasi variational inequality.

REMARK 3. Theorems 3 and 4 were proved first [5] (see also [6] Chap.7) in the case of control constraints set (3.14). In this case Eqs.(4.26) and (4.27) become

(4.37) $\quad u^*(t) = \underline{\mathrm{sgn}}\, p(t) \quad$ a.e. $t \in [o,T^*]$,

respectively

(4.38) $\quad |p(t)|_2 - (Ay(t)+Fy(t),p(t)) = 1 \quad$ a.e. $t \in [o,T^*]$.

REMARK 4. The finite element approximation of the time optimal problem for the system (1.1) leads to a similar problem for the finite dimensional control process

$$y_h' + A_h y_h + \gamma_h(y_h) = u_h \quad \text{a.e. in } R^+$$

$$y_h(o) = y_{o,h}$$

with the control constraints

$$u_h(t) \in U_h \quad \text{a.e. } t > o.$$

Here A_h is a $n(h)\times n(h)$ positive definite matrix, $\gamma_h \subset R^{n(h)} \times R^{n(h)}$ is a maximal monotone graph and U_h is a closed convex and bounded subset of $R^{n(h)}$ containing o as interior point; h is a parameter which tends to zero and $n(h) \to +\infty$ for $h \to 0$.

Most of the above results remain valid in this framework. In particular arguing as in the proofs of Lemma 3 and Theorem 1 we infer that

$$u_h(t) = \ell^*_{h,\varepsilon}(-\partial\varphi_{h,\varepsilon}(y_h(t))), \quad t\geq 0$$

is an optimal feedback law for the corresponding problem (2.1) and $\varphi_{h,\varepsilon}$ (the associated optimal value function) is a solution to the Hamilton-Jacobi equation

$$(A_h y_h + \gamma_h^\varepsilon(y_h), \partial\varphi_{h,\varepsilon}(y_h)) + \ell^*_{h,\varepsilon}(-\partial\varphi_{h,\varepsilon}(y_h)) = g_{h,\varepsilon}(y_h)$$

$$\forall\ y_h \in R^{n(h)}.$$

Here

$$\ell_{h,\varepsilon}(u) = \inf\{\|u-v\|_h/2\ ;\ v\in U_h\}$$

($\|\cdot\|_h$ is the norm of $R^{n(h)}$); γ_h^ε is a C^∞-approximation of γ_h and

$$g_{h,\varepsilon}(y_h) = \pi(\|y_h\|_h^2\ \ \varepsilon^{-1/2}).$$

REFERENCES

1. A.V.Balakrishnan, Applied Functional Analysis, Springer-Varlag 1976.
2. V.Barbu, Nonlinear Semigroups and Differential Equations in Banach Spaces Noordhoff International Publishing Leyden 1976.
3. V.Barbu, Optimal feedback controls for a class of nonlinear distributed parameter systems, SIAM J.Control and Optimiz.21(1983),871-894.
4. V.Barbu, Optimal feedback controls for semilinear parabolic equations, Mathematical Methods in Optimization Cecconi and Zolezzi eds. Lecture Notes in Mathematics, Springer-Varlag 1983.
5. V.Barbu, The time optimal control problem for parabolic variational inequalities, Applied Math. § Optimiz. (to appear).
6. V.Barbu, Optimal Control of Variational Inequalities, Research Notes in Mathematics, Pitman 1984.
7. V.Barbu and T.Precupanu, Convexity and Optimization in Banach Space, Sijthoff § Noordhoff 1978.
8. V.Barbu and G.Da Prato, Hamilton Jacobi Equations in Hilbert Spaces Research Notes in Mathematics 86, Pitman 1983.

9. H.Brézis, <u>Operateurs maximaux monotones et semigroupes de contractions dans les espaces de Hilbert</u> North-Holland 1973.

10. O.Cârjă, On the minimal time function for distributed control systems in Banach spaces, J.O.T.A. (to appear).

11. O.Cărjă, The time optimal control problem for boundary-distributed control systems, Boll.U.M.I. (to appear).

12. F.H.Clarke, Generalized gradients and applications, Adv. in Math.40(1981), 52-67.

13. F.O.Fattorini, The time optimal control problem in Banach space, Applied Math. § Optimiz.(1974), 163-188.

14. O.A.Ladyzhenskaya, V.A.Solonnikov, N.N.Ural'seva, <u>Linear and Quasilinear Equations of Parabolic Type</u>, Amer. Math.Soc.1968.

15. J.L.Lions, <u>Quelques méthodes de résolution des problèmes aux limites non lineaires</u>, Dunod Gauthier-Villars Paris 1969.

16. J.L.Lions, <u>Optimal Control of Systems Governed by Partial Differential Equations</u> Springer Verlag 1971.

17. J.L.Lions, <u>Generalized Solutions of Hamilton-Jacobi</u> Equations Research Notes in Mathematics 69, Pitman 1982.

SOME SINGULAR PERTURBATION PROBLEMS ARISING IN STOCHASTIC CONTROL

A. BENSOUSSAN

INTRODUCTION.

The objective of this article is to present a general approach to treat singular perturbation questions related to stochastic control. The general formulation of the problem is as follows. Let x(t), y(t) be the solution of the system

(1) $$dx = f(x,y,v)dt + \sqrt{2}\, dw \qquad x(0) = x$$

$$\varepsilon dy = g(x,y,v)dt + \sqrt{2\varepsilon}\, db \qquad y(0) = y,$$

where w, b are independent Wiener processes. The parameter ε is small and thus the state (x(t), y(t)) is composed of a slow subsystem x(t) and a fast sub system y(t). The dynamics is controlled by the process v(t). The choice of v(t) is based upon the full observation of x and y. The objective is to minimize the payoff

(2) $$J^{\varepsilon}_{x,y}(v(.)) = E \int_0^{\tau} e^{-\beta t}\, \ell(x^{\varepsilon}(t),y^{\varepsilon}(t))dt$$

where τ denotes the 1st exit time of the process x from the boundary Γ of a smooth bounded domain $\mathcal{O}$.

Let $u_\varepsilon(x,y)$ be defined by

$$u_\varepsilon(x,y) = \operatorname*{Inf}_{v(.)} J^{\varepsilon}_{x,y}(v(.))$$

then u_ε is the solution of the Bellman equation

(3) $$-\Delta_x u_\varepsilon - \frac{1}{\varepsilon}\Delta_y u_\varepsilon + \beta u_\varepsilon = H(x,D_x u_\varepsilon,y,\frac{1}{\varepsilon} D_y u_\varepsilon)$$

$$u_\varepsilon = 0 \qquad \text{for } x \in \Gamma$$

and :

(4) $$H(x,y,p,q) = \operatorname*{Inf}_{v\in U_{ad}} [\ell(x,y,v) + P.f(x,y,v) + q.g(x,y,v)].$$

The question is then the following. What is the behaviour of u_ε as ε tends to 0 ? Does the optimal control problem (1), (2) "converge" towards a simplified optimal control problem, namely one concerning only the slow system. There are several approaches to deal with this problem. The most instructive and general is the method of asymptotic expansions. It consists in writing an expansion of the form :

$$u_\varepsilon(x,y) = u(x) + \varepsilon\, \phi(x,y).$$

Identifying the first order terms we get

$$(5) \qquad - \Delta u - \Delta_y \Phi + \beta u = H(x,D_u,y,D_y\Phi)$$

which we try to match for any pair x,y, by a convenient choice of u and ϕ. In fact (5) must be considered as an equation for ϕ as a function of y and the quantities involving x are constants (with respect to the y variable). The problem for ϕ is an ergodic control problem (note that Φ is defined up to a constant with respect to y). The theory of ergodic control, when applicable shows that there exists one constant (with respect to y) such that the problem (in ϕ)

$$(6) \qquad - \Delta_y \Phi + \chi = H(x, D_u, y, D_y \Phi).$$

is well posed. Note that in (6) the quantities x and Du(x) are parameters, hence in fact $\chi = \chi(x,Du)$. Identifying (5) and (6) yields then the equation

$$(7) \qquad - \Delta u + \beta u = \chi(x,Du)$$

which is the limit problem.

This formal argument shows that the limit problem involves the solution of an intermediary ergodic control problem. This ergodic control problem is related to the fast system.

More precisely, set

$$G(y,v) = g(x,y,v)$$

$$L(y,v) = \ell(x,y,v) + p.f(x,y,v)$$

then the constant χ of (6) can be interpreted as follows

$$(8) \qquad dy = G(y.v)d\tau + \sqrt{2}\, db \qquad\qquad y(0) = y$$

$$K_y(v(.)) = \lim_{T \to \infty} \frac{1}{T} E \int_0^T L(y,v)d\tau$$

and

(9) $$\chi = \inf_{v(.)} K_y(v(.)).$$

This quantity χ is independant of y (but of course as mentionned above depends parametrically on x,p).

Some assumptions are necessary to solve (8), (9).

The main one is that the process $y(\tau)$ be ergodic as $\tau \to \infty$. The simplest case in which this property is fulfilled is when the process $y(\tau)$ remains in a to rus (which is satisfied when g is periodic in y).

This is the case we consider in this article. Other cases of ergodicity are considered in A. BENSOUSSAN - G. BLANKENSHIP [2] and A. BENSOUSSAN [1]. In particular, one can naturally consider cases where the stability conditions of HASMINSKII [3] are verified.

CONTENTS

1. NOTATION - SETTING OF THE PROBLEM.

1.1. *Assumptions*

Let us consider functions

(1.1) $$f(x,y,v) : R^n \times R^d \times U \to R^n$$

$$g(x,y,v) : R^n \times R^d \times U \to R^d$$

$$\ell(x,y,v) : R^n \times R^d \times U \to R$$

continuous and periodic in y with period 1 in each component.

(1.2) U_{ad} compact subset of U (metric space).

Let (Ω,A,P,F^t) be a system in which are constructed two independant standard Wiener processes $b(t)$, $w(t)$ with values in R^d and R^n respectively. We shall define

(1.3)
$$x(t) = x + \sqrt{2}\, w(t)$$
$$y_\varepsilon(t) = y + \sqrt{\tfrac{2}{\varepsilon}}\, b(t).$$

An admissible control is a process $v(t)$ with values in U_{ad}, adapted to F^t. Let us consider the processes

(1.4) $$b_v^\varepsilon(t) = b(t) - \frac{1}{\sqrt{2\varepsilon}} \int_0^t g(x(s),y_\varepsilon(s),v(s))ds$$

(1.5) $$w_v^\varepsilon(t) = w(t) - \frac{1}{\sqrt{2}} \int_0^t f(x(s),y_\varepsilon(s)v(s))ds.$$

Let now $\mathcal{O}$ be a bounded smooth domain of R^n, and $\tau = t_x$ denotes the first exit time of the process $x(t)$ from the domain $\mathcal{O}$. Since we are not going to consider the process $x(t)$ outside $\mathcal{O}$, we may assume without loss of generality that f,g,ℓ are bounded functions. Let us define the probability P^ε (which depends also on the control $v(.)$ and x,y).

(1.6) $$\left.\frac{dP^\varepsilon}{dP}\right|_{F^t} = \exp \{ \int_0^t [\frac{1}{\sqrt{2\varepsilon}}\, g(x(s),y_\varepsilon(s),v(s)).db(s) +$$

$$+ \frac{1}{\sqrt{2}} f(x(s),y_\varepsilon(s),v(s)).dw(s)] - \frac{1}{4}\int_0^t [\frac{1}{\varepsilon} |g(x(s),y\ (s),v(s)|^2 +$$

$$+ |f(x(s),y_\varepsilon(s),v(s))|^2]ds\}.$$

For the system $(\Omega,A,F^t,P\ )$ the processes $b^\varepsilon(t)$ and $w^\varepsilon(t)$ become standard independant Wiener processes and the processes $x(t)$, $y_\varepsilon(t)$ appear as the solutions of

$$(1.7)\qquad dx = f(x(t),y_\varepsilon(t),v(t))dt + \sqrt{2}\, dw_\varepsilon(t)$$

$$dy_\varepsilon = \frac{1}{\varepsilon} g(x(t),y_\varepsilon(t),v(t))dt + \sqrt{\frac{2}{\varepsilon}}\, db_\varepsilon(t)$$

$$x(0) = x \ , \ y_\varepsilon(0) = y.$$

1.2. A problem of stochastic control.

Our objective is to minimize the payoff function $(\beta > 0)$

$$(1.8)\qquad J^\varepsilon_{x,y}(v(.)) = E^\varepsilon \int_0^{\tau_x} \ell(x(t),y_\varepsilon(t),v(t))e^{-\beta t}dt.$$

If we set

$$(1.9)\qquad u_\varepsilon(x,y) = \operatorname*{Inf}_{v(.)} J^\varepsilon_{x,y}(v(.))$$

Then u_ε is the unique solution of the H.J.B. (Hamilton Jacobi Bellman) equation

$$(1.10)\qquad -\Delta_x u_\varepsilon - \frac{1}{\varepsilon}\Delta_y u_\varepsilon + \beta u_\varepsilon = H(x,y,D_x u_\varepsilon,\frac{1}{\varepsilon} D_y u_\varepsilon)$$

$u_\varepsilon = 0$ for $x \in \Gamma$, $\forall y$

u_ε periodic in y

$u_\varepsilon \in W^{2,p}(\mathcal{O} \times Y)$ $\quad 2 \le p < \infty$

where $\Gamma = \partial\mathcal{O}$ is the boundary of $\mathcal{O}$ and where

$$(1.11)\qquad H(x,y,p,q) = \operatorname*{Inf}_{v\in U_{ad}} [\ell(x,y,v) + p.f(x,y,v) + q.g(x,u,v)]$$

$$= \operatorname*{Inf}_{v\in U_{ad}} L(x,y,p,q,v).$$

Moreover there exists a Borel map $\hat{V}(x,y,p,q)$ with values in U_{ad}, such that

$$H(x,y,p,q) = L(x,y,p,q,\hat{V}) \tag{1.12}$$

We can define an optimal feedback for (1.8) by setting

$$v_\varepsilon(x,y) = V(x,y,D_x u_\varepsilon, D_y u_\varepsilon) \tag{1.13}$$

and the process

$$v_\varepsilon(t) = v_\varepsilon(x(t),y(t)) \tag{1.14}$$

is an optimal control for (1.8).

Our objective is to study the behaviour of u_ε as ε tends to 0.

2. *LIMIT PROBLEM.*

2.1. *Notation.*

Let $v(y)$ be any Borel function with values in U_{ad}. For such a $v(.)$ and a given x, let us define $m^v(x,y)$, which is the solution of the problem

$$- \Delta_y m + \mathrm{div}_y(m\, g(x,y,v(y))) = 0 \tag{2.1}$$

$$m \in H^1(Y), \quad m \text{ periodic},$$

where $Y =]0,1[^n$

For x,p fixed we can consider the quantity

$$\chi(x,p) = \mathrm{Inf} \int_Y m^v(x,y)(\ell(x,y,v(y)) + p.f(x,y,v(y)))dy \tag{2.2}$$

where the infimum runs over all Borel functions $v(.)$.

The function χ is u.s.c., uniformly Lipschitz in p, with linear growth. Therefore, considering the problem

$$- \Delta u + \beta u = \chi(x,Du) \quad , \quad u|_\Gamma = 0 \tag{2.3}$$

it has a unique solution in $W^{2,p}(\mathcal{O})$, $2 \le p < \infty$.

2.2. Statement of the main result.

The main result of this article is the following

Theorem 2.1 : Assume (1.1) and (1.2). Then one has

$$u_\varepsilon \to u \quad \text{in } H^1(\mathcal{O} \times Y) \quad \text{strongly} \tag{2.4}$$

Remark 2.1. We do not give any estimate of the rate of convergence. This is due to the fact that we have minimal assumptions. Under further regularity assumptions, it is possible to derive an estimate of the rate of convergence.

Remark 2.2. There are two techniques to prove the result (2.4). One is based on the method of asymptotic expansions, whose principle has been explained in the introduction. Suppose that one can solve the equation (6) of the introduction and set

$$\tilde{u}_\varepsilon = u_\varepsilon - u - \varepsilon\, \phi$$

One can derive an equation for $\tilde{u}_\varepsilon$. By maximum principle considerations, it is possible to show that

$$\|\tilde{u}_\varepsilon\|_{L^\infty} \le C\, \varepsilon$$

This method has the advantage of giving an estimate of the rate of convergence. But it has the serious drawback of requiring a lot of regularity.

The other method which requires only the assumptions (1.1) and (1.2) is based upon a priori estimates and energy type techniques.

3. *PROOF OF THEOREM 2.1.*

3.1. A priori estimates

Let us consider the problem

$$- \varepsilon\, \Delta_x m_\varepsilon - \Delta_y m_\varepsilon + \operatorname{div}_y(m_\varepsilon\, g(x,y,v_\varepsilon)) = 0 \tag{3.1}$$

$$\left.\frac{\partial m_\varepsilon}{\partial \nu}\right|_\Gamma = 0 \quad , \; m_\varepsilon \text{ periodic in } y$$

$$m_\varepsilon \in H^1(\mathcal{O} \times Y).$$

This problem has one and only one solution. Moreover the following estimate holds

(3.2) $$0 < \delta \le m_\varepsilon(x,y) \le \delta_1$$

where δ, δ_1 are positive constants, which do not depend of ε. With this result it is possible to prove the

Lemma 3.1. The following estimates hold

(3.3) $$\| u_\varepsilon \|_{L^\infty} \le C, \qquad |D_x u_\varepsilon|_{L^2} \le C$$

$$|D_y u_\varepsilon|_{L^2} \le C \sqrt{\varepsilon}$$

Proof.

Using the definition of v_ε (cf (1.13)), one can write (1.10) as follows

(3.4) $$- \Delta_x u_\varepsilon - \frac{1}{\varepsilon} \Delta_y u_\varepsilon + \beta u_\varepsilon = \ell(x,y,v_\varepsilon) + Du_\varepsilon . f(x,y,v_\varepsilon) + \frac{1}{\varepsilon} D_y u_\varepsilon . g(x,y,v_\varepsilon).$$

Multiply (3.4) by $m_\varepsilon u_\varepsilon$ and integrate. Using the boundary conditions we deduce

(3.5) $$\iint m_\varepsilon |D_x u_\varepsilon|^2 dxdy + \frac{1}{\varepsilon} \iint m_\varepsilon |D_y u_\varepsilon|^2 dxdy + \beta \iint m_\varepsilon u_\varepsilon^2 dxdy = \iint (\ell_\varepsilon + f_\varepsilon . D_x u_\varepsilon) m_\varepsilon u_\varepsilon \, dxdy.$$

Note that from the maximum principle, we deduce from (3.4) that

(3.6) $$\| u_\varepsilon \|_{L^\infty} \le \frac{\| \ell \|}{\beta}$$

By virtue of (3.2) the remainder of the estimates (3.3) follow from the equality (3.5)

□

Lemma 3.2. Let $(x) \in H_0^1(\mathcal{O}) \cap H^2(\mathcal{O})$ then the following relation holds

(3.7) $$\iint m_\varepsilon |D_x(u_\varepsilon - \phi)|^2 dxdy + \frac{1}{\varepsilon} \iint m_\varepsilon |D_y u_\varepsilon|^2 dxdy + \iint \beta m_\varepsilon (u_\varepsilon - \phi)^2 dxdy =$$
$$= \iint m_\varepsilon u_\varepsilon (\Delta\phi - \beta\phi) dxdy + \int |D_x \phi|^2 dx + \int \beta^2 \phi^2 dx + \iint (\ell_\varepsilon + f_\varepsilon D_x u_\varepsilon) m_\varepsilon (u_\varepsilon - \phi) dxdy$$

Proof.

Multiply (3.4) by $m_\varepsilon \phi$ and integrate. Similarly multiply (3.1) by $u_\varepsilon \phi$ and integrate. Use (3.5) and expand (3.7). By simple calculations the desired expression (3.7) obtains.

□

3.2. *Convergence.*

Let us first prove the

Lemma 3.3. Let us consider a subsequence of u_ε such that

(3.8) $$u_\varepsilon \to u \quad \text{in } H^1(\mathcal{O} \times Y) \text{ weakly.}$$

Then u is a function of x only, belongs to $H_0^1(\mathcal{O})$. Moreover the convergence (3.8) is strong.

Proof.

From Lemma 3.1, we know that subsequences for which the property (3.8) holds exist. From the 3rd estimate (3.3) and

$$|D_y u|^2 \le \underline{\lim}\, |D_y u_\varepsilon|^2 = 0$$

we deduce that $u = u(x)$. Clearly $u \in H_0^1(\mathcal{O})$, since u vanishes for x in Γ. We have

$$X_\varepsilon = \iint m_\varepsilon |D_x(u_\varepsilon - u)|^2 dxdy + \frac{1}{\varepsilon}\iint m_\varepsilon |D_y u_\varepsilon|^2\, dxdy +$$

$$+ \beta \iint m_\varepsilon (u_\varepsilon - u)^2\, dxdy \le 2 \iint m_\varepsilon |D_x(u_\varepsilon - \phi_k)|^2\, dxdy +$$

$$+ \frac{2}{\varepsilon} \iint m_\varepsilon |D_y u_\varepsilon|^2\, dxdy + 2\beta \iint m_\varepsilon (u_\varepsilon - \phi_k)^2\, dxdy +$$

$$+ 2 \int |D_x(\phi_k - u)|^2\, dx + 2\beta \int (\phi_k - u)^2\, dx$$

and from (3.7) with $\phi = \phi_k$ we deduce

$$X_\varepsilon \leqq 2 \iint m_\varepsilon u_\varepsilon (\Delta\phi_k - \beta\phi_k) dxdy + 2 \int |D_x \phi_k|^2 ds + 2\beta \int \phi_k^2\, dx +$$

$$+ 2 \iint (\ell_\varepsilon + f_\varepsilon\, D_x\, u_\varepsilon) m_\varepsilon (u_\varepsilon - \phi_k) dxdy + 2 \int |D_x(\phi_k - u)|^2 dx +$$

$$+ 2\beta \int (\phi_k - u)^2 dx.$$

We can with loss of generality assume that $m_\varepsilon \to m^*$ in $L^\infty(\mathcal{O} \times Y)$ weak star and $m_\varepsilon(\ell_\varepsilon + f_\varepsilon D_x u_\varepsilon) \to \eta$ in $L^2(\mathcal{O} \times Y)$ weakly. Note that

$$\int_Y m_\varepsilon(x,y)dy = 1 \quad \text{implies} \quad \int_Y m^*(x,y)dy = 1.$$

We then deduce that

$$\overline{\lim}\, X_\varepsilon \le 2 \int u(\Delta\phi_k - \beta\phi_k)dx + 2 \int |D_x \phi_k|^2 dx + 2\beta \int \phi_k^2 dx +$$
$$+ 2 \iint \eta(u - \phi_k)dxdy + 2 \int |D_x(\phi_k - u)|^2 dx + 2\beta \int (\phi_k - u)^2 dx.$$

Letting next k tend to $+\infty$, we obtain

$$\overline{\lim}\, X_\varepsilon = 0.$$

Since clearly

$$X_\varepsilon \ge \delta \iint |D_x(u_\varepsilon - u)|^2 dxdy + \beta\delta \iint (u_\varepsilon - u)^2 dxdy$$

the strong convergence property is established.

□

The next step is to identify the limit. We need the following intermediary result. Let $F(x,y,v) : \mathcal{O} \times Y \times U_{ad} \to R$ such that

(3.9) F is periodic in y, $|F| \le \bar{F}(x) \in L^2(\mathcal{O})$:

F is measurable, and continuous in y,v.

Define an Hamiltonian by setting

$$\mathcal{H}(x,y,q) = \inf_{v \in U_{ad}} [F(x,y,v) + q.g(x,y,v)]. \tag{3.10}$$

Consider next the problem : to find a pair $\Phi_\varepsilon(x,y)$ and Λ_ε a scalar such that

(3.11) $$-\varepsilon\, \Delta_x \Phi_\varepsilon - \Delta_y \Phi_\varepsilon = \mathcal{H}(x,y,D_y\Phi_\varepsilon) - \Lambda_\varepsilon$$

$$\frac{\partial\Phi_\varepsilon}{\partial\nu}\Big|_\Gamma = 0 \quad , \quad \Phi_\varepsilon \text{ periodic in } y$$

$$\Phi_\varepsilon \in W^{2,p}(\mathcal{O} \times Y).$$

On the other hand, consider for x frozen, the problem

(3.12) $$- \Delta_y \Phi = \mathcal{H}(x,y,D_y\Phi) - \wedge(x)$$

Φ periodic in y , $\Phi \in W^{2,p}(Y)$ $\forall x$.

The following result holds

Lemma 3.4. We have

(3.13) $$|\mathcal{O}| \wedge_\varepsilon \to \int_{\mathcal{O}} \wedge(x)dx, \text{ as } \varepsilon \to 0.$$ □

We can then proceed with the proof of Theorem 2.1. Let u_ε be a subsequence which converges to some u in $H^1(\mathcal{O} \times Y)$ weakly. By Lemma 3.3, we can assert that u does not depend on y and that the convergence is strong. Moreover $u \in H_0^1(\mathcal{O})$.
Let $\phi \in C_0^\infty(\mathcal{O})$, $\phi \geq 0$. We deduce from (1.10), multiplying by ϕm_ε that

(3.14) $$2 \iint D_x u_\varepsilon D_x \phi m_\varepsilon \, dxdy + \iint m_\varepsilon u_\varepsilon \Delta\phi \, dxdy + \beta \iint m_\varepsilon u_\varepsilon \phi \, dxdy =$$
$$= \iint (\ell(x,y,v_\varepsilon) + D_x u.f(x,y,v_\varepsilon))m_\varepsilon \phi \, dxdy + \iint D_x(u_\varepsilon - u).f(x,y,v_\varepsilon)m_\varepsilon \phi dxdy.$$

Consider

$$F(x,y,v) = \phi(x) \; (\ell(x,y,v) + Du.f(x,y,v)).$$

We want to apply (3.11), (3.12). From the interpretation of $\wedge_\varepsilon$ (cf by analogy (2.2)), we can assert that

(3.15) $$\iint (\ell(x,y,v_\varepsilon) + D_x u.f(x,y,v_\varepsilon))m_\varepsilon \phi \, dxdy \geq |\mathcal{O}| \wedge_\varepsilon .$$

For this problem, the function $\wedge(x)$ entering into (3.12) is given by

$$\wedge(x) = \chi(x,Du)\phi(x)$$

hence from Lemma 3.4, we have

(3.16) $$|\mathcal{O}|\wedge_\varepsilon \to \int \chi(x,Du) \; \phi(x)dx .$$

We can also assume that $m_\varepsilon \to m^*$ in L^∞ weak star, and $\int_Y m^*(x,y)dy = 1$, a.e.
From (3.14) we get

(3.17) $$2\iint D_x u_\varepsilon . D_x \; m_\varepsilon \, dxdy + \iint m_\varepsilon u_\varepsilon \, \Delta\phi \, dxdy +$$

$$+ \beta \iint m_\varepsilon u_\varepsilon \phi \, dxdy \geq |\mathcal{O}| \wedge_\varepsilon + \iint D_x(u_\varepsilon - u) . f(x,y,v_\varepsilon) m_\varepsilon \; \phi \; dxdy$$

and since $u_\varepsilon \to u$ in H^1 strongly, we can pass to the limit in the inequality (3.17). This yields clearly, recalling that u does not depend on y

$$2\int D_x \, u \, D_x \, \phi \; dx + \int u \, \Delta\phi \; dx + \beta \int u \, \phi \; dx$$

$$\geq \int \chi(x,Du)\phi(x)\,dx$$

hence

$$\int (-\Delta u + \beta u)\,\phi\, dx \geq \int \chi(x,Du)\phi(x)\,dx.$$

Since ϕ is arbitrary non negative, it follows

$$-\Delta u + \beta u \geq \chi(x,Du).$$

A reverse inequality is proved in a similar and actually simpler way. This completes the proof of the desired result.

<u>REFERENCES</u>

[1] A. BENSOUSSAN, <u>Perturbation Methods in Optimal Control</u>, Book in preparation

[2] A. BENSOUSSAN, G. BLANKENSHIP, <u>Singular Perturbations in Stochastic Control</u>, in P. KOKOTOVIC, A. BENSOUSSAN, G. BLANKENSHIP, ed, lecture Notes Springer Verlag, to be published

[3] R.Z. HASMINSKII, <u>Stochastic Stability of Differential Equations</u>, Sijthoff - Noordhoff, 1980

SOME RESULTS ON STATIONARY BELLMAN EQUATION IN HILBERT SPACES

G. DA PRATO
Scuola Normale Superiore
56100 PISA, Italy

1. INTRODUCTION

We shall be concerned here with the problem

$$\begin{cases} \phi_t - \frac{\varepsilon}{2} \operatorname{Tr}(\phi_{xx}(Bx)S(Bx)^\star) + \frac{1}{2} |\phi_x|^2 - \langle Ax, \phi_x \rangle = g \\ \phi(0,x) = \phi_0(x) \quad , \quad \varepsilon > 0 \end{cases} \tag{1.1}$$

as well as with the stationary equation

$$\frac{\varepsilon}{2} \operatorname{Tr}(\phi_{xx}(Bx)S(Bx)^\star) + \frac{1}{2} |\phi_x|^2 - \langle Ax, \phi_x \rangle = g \tag{1.2}$$

Here g and ϕ_0 are convex mappings from a Hilbert space H to $\mathbb{R}$, A is a linear operator (generally unbounded) in H, S is a positive nuclear operator in a Hilbert space X and $B \in \mathcal{L}(X,H)$. Moreover $(Bx)^\star$ denotes the adjoint of Bx and we look for a solution ϕ convex in x. Eq. (1.2) is connected with the following optimal control problem:

minimize

$$J(x,u) = E \int_0^\infty [g(y(s)) + \frac{1}{2} |u(s)|^2] ds \tag{1.3}$$

over all $u \in M_w^2(0,\infty;H)$ subject to the state equation

$$\begin{cases} dy = (Ay + u)dt + \sqrt{\varepsilon} Bu \, dw_t \\ y(0) = x \quad . \end{cases} \tag{1.4}$$

Here w_t is a X-valued Brownian motion whose covariance is S, and $M_w^2(0,T,H)$ denotes the set of all square integrable H-valued stochastic processes that are adapted to w. Our goal is to prove that, under suitable conditions, the solution of Eq. (1.1) converges to a generalized solution of Eq. (1.2).

Section 2 is devoted to notations and hypotheses, in Section 3 we report, with some improvement, the results of [1] and [2] concerning Eq. (1.1).

In Section 4 we study Eq. (1.2) and in Section 5 we solve the control problem using Dynamic Programming.

2. NOTATIONS AND HYPOTHESES

We shall denote by H and X real Hilbert spaces and by K (resp. K^+) the set of all convex (resp. convex and non-negative) mappings from H to $\mathbb{R}$ that are continuous. We shall denote by $\mathbb{N}$ the set of all non-negative integers.

Regularization of a convex function

For any $\phi \in K$ we set:

$$\phi_\alpha(x) = \min\left\{\phi(y) + \frac{|x-y|^2}{2\alpha} \; ; \; y \in H\right\}, \; \alpha > 0 \; , \; x \in H \; . \qquad (2.1)$$

We remark that the minimum is reached at $x = x_\alpha$ where:

$$x_\alpha = (1 + \alpha F)^{-1}(x) \qquad (2.2)$$

and $F = \partial\phi$ is the subdifferential of ϕ.
As is well known, F is maximal monotone. Thus we can write

$$\phi_\alpha(x) = \phi(x_\alpha) + \frac{\alpha}{2}\left|\frac{x - x_\alpha}{\alpha}\right|^2 = \phi(x_\alpha) + \frac{\alpha}{2}|F_\alpha(x)|^2 \qquad (2.3)$$

where

$$F_\alpha(x) = \frac{x - x_\alpha}{\alpha} \; . \qquad (2.4)$$

The following lemma collects, for further use, some properties of ϕ_α and F_α (see for instance [3]).

<u>Lemma 2.1</u>

<u>Let</u> $\phi \in K$, <u>then for any</u> $\alpha > 0$, ϕ_α <u>is differentiable and</u>:

$$|x_\alpha - y_\alpha| \leq |x - y| \quad , \quad \forall x,y \in H \qquad (2.5)$$

$$\phi'_\alpha(x) = \frac{d\phi_\alpha}{dx}(x) = F_\alpha(x) \in F(x_\alpha) \qquad (2.6)$$

$$|F_\alpha(x)| \leq |F(x)| = \text{Sup}\,\{|\eta| \; ; \eta \in F(x)|\} \qquad (2.7)$$

$$\phi(x_\alpha) \leqslant \phi_\alpha(x) \leqslant \phi(x) \quad , \quad \forall x \in H \tag{2.8}$$

$$0 \leqslant \phi(x) - \phi_\alpha(x) \leqslant \frac{\alpha}{2} |F_\alpha(x)|^2 \tag{2.9}$$

<u>Lemma 2.2</u>

<u>If</u> $\phi \in K$, <u>we have</u>:

$$|x_\alpha| \leqslant |x| + 2\alpha \sup_{|y| \leqslant 1} |\phi(y)| \tag{2.10}$$

<u>Proof</u>

Since F is monotone, we have

$$|x_\alpha|^2 = \langle x, x_\alpha \rangle - \alpha \langle F_\alpha(x), x_\alpha \rangle \leqslant \langle x, x_\alpha \rangle - \alpha \langle \eta, x_\alpha \rangle \quad \forall \eta \in F(0)$$

from which

$$|x_\alpha| \leqslant |x| + \alpha |\eta| \qquad \forall \eta \in F(0) .$$

It suffices now to notice that, by

$$\phi(z) - \phi(0) \geqslant \langle \eta, z \rangle \quad , \quad \forall \eta \in B_1 = \{z \in H \, ; |z| \leqslant 1 \}$$

it follows that

$$|\eta| \leqslant |\phi(0)| + \sup_{|z| \leqslant 1} |\phi(z)| \qquad \#$$

In the sequel we shall use (as in [1]) the device to approximate $\frac{1}{2} |\phi_x|^2$ by $\dfrac{\phi - \phi_\alpha}{\alpha}$; let us denote by $R_{\alpha,\phi}$ the error:

$$R_{\alpha,\phi}(x) = \frac{1}{\alpha} (\phi(x) - \phi_\alpha(x)) - \frac{1}{2} |\phi'(x)|^2 \quad ; \tag{2.11}$$

by (2.9), we have

<u>Lemma 2.3</u>

<u>Let</u> $\phi \in K$, <u>then we have</u>

$$|R_{\alpha,\phi}(x)| \leqslant \frac{1}{2} (|F(x)|^2 - |F_\alpha(x)|^2) \quad . \tag{2.12}$$

The following lemma is proved in [1] .

<u>Lemma 2.4</u>

<u>Let</u> $\phi,\overline{\phi}\in K$, $F=\partial\phi$, $\overline{F}=\partial\overline{\phi}$; $x_\alpha=(1+\alpha F)^{-1}(x)$, $\overline{x}_\alpha=(1+\alpha\overline{F})^{-1}(x)$. <u>Then</u>

$$\phi_\alpha(x)-\overline{\phi}_\alpha(x)\leqslant\phi(\overline{x}_\alpha)-\overline{\phi}(\overline{x}_\alpha)\ . \tag{2.13}$$

<u>If, moreover,</u> ϕ <u>and</u> $\overline{\phi}$ <u>are differentiable, we have</u>

$$|F_\alpha(x)-\overline{F}_\alpha(x)|\leqslant|F(x_\alpha)-\overline{F}(x_\alpha)|\ . \tag{2.14}$$

<u>Some semi-norms</u>

Let $\phi:H\to\mathbb{R}$ be h times continuously differentiable. We set:

$$|\phi|_{h,n}=\sup_{x\in H}\frac{|\phi^{(h)}(x)|}{1+|x|^n}\ ,\quad n\in\mathbb{N} \tag{2.15}$$

and

$$\|\phi\|_{h,n}=\sup_{\substack{x,y\in H\\ x\neq y}}\frac{|\phi^{(h)}(x)-\phi^{(h)}(y)|}{|x-y|\,(1+\sup(|x|^n,|y|^n))}\ . \tag{2.16}$$

If n_i are integers such that $n_1\geqslant n_2\geqslant\dots\geqslant n_h\geqslant 0$, we shall set

$$C^h(H;n_0,n_1,\dots,n_h)=\{\phi:|\phi|_{C^h}=\sum_{i=0}^{h}|\phi|_{i,n_i}<\infty\} \tag{2.17}$$

and

$$C^h_{Lip}(H;n_0,n_1,\dots,n_{h+1})=\{\phi:\|\phi\|_{C^h_{Lip}}=\sum_{i=0}^{h}|\phi|_{i,n_i}+\|\phi\|_{h,n_{h+1}}<\infty\} \tag{2.18}$$

$C^h(H;n_0,n_1,\dots,n_h)$ (resp. $C^h_{Lip}(H;n_0,n_1,\dots,n_{h+1})$), endowed with the norm $|\ |_{C^h}$ (resp. $\|\ \|_{C^h_{Lip}}$) are Banach spaces.

We shall denote by $C([0,T];C^h(H;n_0,n_1,\dots,n_h))$ (resp. $C([0,T];C^h_{Lip}(H;n_0,n_1,\dots,n_{h+1}))$ the set of all continuous mappings $\phi:[0,T]\times H\to\mathbb{R}$ such that

$$\sup_{t\in[0,T]}|\phi(t,\cdot)|_{C^h}<\infty\ \text{(resp.}\ \sup_{t\in[0,T]}\|\phi(t,\cdot)\|_{C^h_{Lip}}<\infty\text{)}\ . \tag{2.19}$$

Lemma 2.5

Let $n,m \in \mathbb{N}$, $\phi \in C(H;n) \cap K$. Then there exists a positive increasing function C_m (non depending on n) such that:

$$\frac{1 + \operatorname{Sup}(|x_\alpha|,|y_\alpha|)^m}{1 + \operatorname{Sup}(|x|,|y|)^m} \leqslant 1 + \alpha C_m(|\phi|_{0,n}) \quad , \quad \alpha \in [0,1] \tag{2.20}$$

Proof

By (2.10) it follows that $|x_\alpha| \leqslant |x| + 4\alpha\,|\phi|_{0,n}$ which implies (2.20) #

Lemma 2.6

Let $\phi \in C(H;n) \cap K^+$, then

$$|\phi_\alpha|_{0,n} \leqslant |\phi|_{0,n} \qquad \forall n \in \mathbb{N}\ ,\ \alpha > 0 \quad . \tag{2.21}$$

If, moreover, ϕ is differentiable, we have

$$|\phi_\alpha|_{1,n} \leqslant |\phi|_{1,n} \qquad \forall n \in \mathbb{N}\ ,\ \alpha > 0 \tag{2.22}$$

$$\|\phi_\alpha\|_{1,m} \leqslant (1 + \alpha\, C_m(|\phi|_{0,n}))\|\phi\|_{1,m} \quad , \quad \alpha \in [0,1],\ m,n \in \mathbb{N} \ . \tag{2.23}$$

Proof

(2.21) (resp. (2.22)) follows from (2.8) (resp. (2.7)). Furthermore we have

$$|\phi_\alpha'(x) - \phi_\alpha'(y)| = |F(x_\alpha) - F(y_\alpha)| \leqslant \|\phi\|_{1,n} \frac{1 + \operatorname{Sup}(|x_\alpha|,|y_\alpha|)^n}{1 + \operatorname{Sup}(|x|,|y|)^n}$$

thus (2.20) implies (2.23) #

Lemma 2.7

Assume that $\phi,\bar{\phi} \in C(H;n) \cap K^+$ and let $\rho = \operatorname{Sup}(|\phi|_{0,n},|\bar{\phi}|_{0,n})$. Then

$$|\phi_\alpha - \bar{\phi}_\alpha|_{0,n} \leqslant (1 + \alpha\, C_n(\rho))\ |\phi - \bar{\phi}|_{0,n} \tag{2.24}$$

and if $\phi,\bar{\phi}$ are differentiable

$$|\phi_\alpha - \bar{\phi}_\alpha|_{1,n} \leqslant (1 + \alpha\, C_n(\rho))|\phi - \bar{\phi}|_{1,n} \quad . \tag{2.25}$$

Proof

Follows from (2.13), (2.14) and (2.20) #

Lemma 2.8

Assume that $\phi \in C(H;n) \cap K^+$. *Then, for any* $\alpha > 0$ *we have:*

$$|\phi_\alpha|_{0,2} \leq \frac{1}{2\alpha} \operatorname{Sup}\{1, 2\alpha|\phi|_{0,n}\} \tag{2.26}$$

$$|\phi_\alpha|_{1,1} \leq \frac{2}{\alpha} \operatorname{Sup}\{1, 2\alpha|\phi|_{0,n}\} \tag{2.27}$$

$$\|\phi_\alpha\|_{1,0} \leq \frac{1}{\alpha} . \tag{2.28}$$

Moreover

$$\lim_{\alpha \to 0} |\phi - \phi_\alpha|_{0,n} = 0 . \tag{2.29}$$

Proof

We have

$$\phi_\alpha(x) \leq \phi(0) + \frac{1}{2\alpha}|x|^2 \leq |\phi|_{1,n} + \frac{1}{2\alpha}|x|^2$$

and (2.26) follows. (2.27) and (2.28) are easy.

Concerning (2.29) we notice that, by (2.9) it follows that

$$|\phi(x) - \phi_\alpha(x)| \leq \frac{\alpha}{2}|F(x)|^2 . \tag{2.30}$$

Since

$$\phi(x+z) - \phi(x) \geq \langle F(x), z \rangle \qquad \forall |z| \leq 1$$

we have

$$|F(x)| \leq |\phi(x)| + \operatorname{Sup}\{|\phi(x+z)|, |z| \leq 1\}$$

thus, by plugging in (2.30)

$$|\phi(x) - \phi_\alpha(x)| \leq \frac{\alpha}{2}|\phi|_{1,n}(2 + |x|^n + (1+|x|)^n)$$

which implies (2.29) #

Lemma 2.9

Let $\phi \in C^1_{Lip}(H;n,m,p)$ *with* $n \geq 2m+p$. *Then there exists an increasing positive function* $D = D_{n,m,p}$ *such that:*

$$|R_{\alpha,\phi}|_{0,n} \leqslant \alpha\, D(|\phi|_{0,n})\; |\phi|^2_{1,m}\; \|\phi\|_{1,p} \quad . \tag{2.31}$$

Proof

By (2.12) we have

$$\begin{aligned}|R_{\alpha,\phi}(x)| &\leqslant |F(x)||F(x) - F(x_\alpha)| \\ &\leqslant (1 + |x|^m)(1 + \operatorname{Sup}(|x|^p, |x_\alpha|^p))|x - x_\alpha||\phi|_{1,m}\|\phi\|_{1,p} \\ &\leqslant \alpha(1 + |x|^m)(1 + \operatorname{Sup}(|x|^p, |x_\alpha|^p)|\phi|^2_{1,m}\;\|\phi\|_{1,p}\end{aligned}$$

and (2.31) follows #

The linear problem

We consider here the linear problem:

$$\begin{cases} \psi_t = \frac{\varepsilon}{2}\operatorname{Tr}(S(Bx)\psi_{xx}(Bx)^\star) - \langle Ax, \psi_x\rangle + g \\ \psi(0,x) = \psi_0(x) \end{cases} \tag{2.32}$$

under the following hypotheses:

a) A is the infinitesimal generator of a strongly continuous semi-group e^{tA} in H.

b) $B \in \mathcal{L}(H;\mathcal{L}(X,H))$.

c) w_t is an X-valued Brownian motion whose covariance S is given by

$$Sx = \sum_{i=1}^{\infty} \lambda_i \langle x, e_i\rangle e_i \quad , \qquad \lambda_i \geqslant 0 \tag{2.33}$$

and $\{e_i\}$ is an orthonormal basis in H.

d) There exists $\omega \in \mathbb{R}$ such that

$$2\langle Ax, x\rangle + \operatorname{Tr}(S(Bx)(Bx)^\star) \leqslant \omega|x|^2 \quad , \qquad \forall x \in D(A).$$

We remark that hypothesis (2.33)-b can be weakened and B can be

taken unbounded (see [4]).
Along with (2.32) we shall consider the finite-dimensional approximating problem:

$$\begin{cases} \psi_t^n = \frac{\varepsilon}{2} \operatorname{Tr}(S_n(B_nP_nx)\,\psi_{xx}^n(B_nP_nx)^*) - \langle A_nx_n, \psi_x^n\rangle + g(x_n) \\ \psi_n(0,x) = \psi_0(P_nx) \end{cases} \tag{2.34}$$

where

$$\begin{cases} P_nx = \sum_{i=1}^{n} \lambda_i \langle x, e_i\rangle e_i \quad , \quad S_n = SP_n,\ A_n = P_nAP_n \\ B_n = P_nBP_n \end{cases} \tag{2.35}$$

and we make the following additional assumption

$$\begin{cases} \text{a)} \quad \{e_n\} \subset D(A) \\ \text{b)} \quad \lim_{n\to\infty} e^{tA_n}x = e^{tA}x \quad , \quad \text{uniformly on the compact} \\ \qquad \text{subset of } [0,\infty[\ . \end{cases} \tag{2.36}$$

To solve Eq. (2.32) we consider the following stochastic differential equation:

$$\begin{cases} d\zeta = A\zeta ds + \sqrt{\varepsilon}\, B\zeta\, dw_s \\ \zeta(t) = x \end{cases} \tag{2.37}$$

whose solution we denote by $\zeta(s,t,x)$.

The following lemma is a consequence of standard results on abstract stochastic equations and estimates on Itô integral.

<u>Lemma 2.10</u> <u>Under hypotheses</u> (2.33), Eq. (2.37) <u>has a unique mild solution</u> $\zeta(s,t,x)$. <u>Moreover for any</u> $n \in \mathbb{N}$ <u>there exists</u> ω_n <u>such that</u>

$$|E\,\psi(\zeta(s,t,\cdot)|_{k,n} \leqslant e^{\omega_n(s-t)}|\psi|_{k,n} \ . \tag{2.38}$$

Now we shall write the mild solution of Eq. (2.32) as

$$\psi(t,x) = E\,\psi_0(\zeta(t,0,x)) + E\int_0^t g(\zeta(t,s,x)\,ds \tag{2.39}$$

Along with (2.37) we shall consider the approximating equation

$$\begin{cases} d\zeta_n = A_n\zeta_n \, ds + \sqrt{\varepsilon}\, B_n\zeta_n \, dw_s^n \quad , \qquad w_s^n = P^n w_s \\ \zeta_n(t) = x \quad . \end{cases} \tag{2.40}$$

The mild solution of (2.34) is given by

$$\psi^n(t,x) = E\,\psi_0(\zeta_n(t,0,P_n x)) + E\int_0^t g(\zeta_n(t,s,P_n x))ds \quad . \tag{2.41}$$

Remark that if ψ_0 and g are twice continuously differentiable, then ψ^n is a classical solution of Eq. (2.34).

<u>Lemma 2.11</u>

<u>Assume</u> (2.33) <u>and</u> (2.36), <u>let</u> ψ (resp. ψ^n) <u>be given by</u> (2.39) (resp. (2.41)). <u>Then for any</u> $x \in H$, <u>we have</u>

$$\lim_{n\to\infty} \psi^n(t,x) = \psi(t,x) \tag{2.42}$$

<u>uniformly in</u> t <u>in any compact subset of</u> $[0,+\infty[$.

<u>Proof</u>

The proof involves a martingale type inequality for stochastic convolution (see [7]) and standard arguments #

3. BELLMAN EVOLUTION EQUATION

We assume here (2.33) and (2.36) and consider the equation

$$\begin{cases} \phi_t = \frac{1}{2}\,\varepsilon\,\mathrm{Tr}(S(Bx)\phi_{xx}(Bx)^\star) - \frac{1}{2}|\phi_x|^2 + \langle Ax,\phi_x\rangle + g \\ \phi(0,x) = \phi_0(x) \end{cases} \tag{3.1}$$

We consider also the following approximating equations:

$$\begin{cases} \phi_t^\alpha = \frac{1}{2}\,\varepsilon\,\mathrm{Tr}(S(Bx)\phi_{xx}^\alpha(Bx^\star)) - \frac{1}{\alpha}\,(\phi^\alpha - \phi_\alpha^\alpha) + \langle Ax,\phi_x^\alpha\rangle + g \\ \phi^\alpha(0,x) = \phi_0(x) \end{cases} \tag{3.2}$$

and

$$\begin{cases} \phi_t^{\alpha,i} = \frac{1}{2}\,\varepsilon\,\mathrm{Tr}(S_i(B_i x)\phi_{xx}^{\alpha,i}(B_i x)^*) - \frac{1}{\alpha}(\phi^{\alpha,i} - \phi_\alpha^{\alpha,i}) + \\ \quad + < A_i x, \phi_x^{\alpha,i} > + g(P_i x) \\ \phi^{\alpha,i}(0,x) = \phi(P_i x) \quad . \end{cases} \tag{3.3}$$

We shall write Eq. (3.2) and (3.3) in the following integral form:

$$\phi^\alpha(t,x) = e^{-t/\alpha}\, E\phi_0(\zeta(t,0,x)) + \int_0^t e^{-(t-s)/\alpha}\, E(\frac{\phi_\alpha^\alpha}{\alpha} + g)(\zeta(t,s,x))ds \tag{3.4}$$

$$\phi^{\alpha,i}(t,x) = e^{-t/\alpha} E\phi_0(\zeta_i(t,0,P_i x)) + \int_0^t e^{-(t-s)/\alpha} E(\frac{\phi_\alpha^{\alpha,i}}{\alpha} + g)(\zeta_i(t,s,P_i x))ds \tag{3.5}$$

When ϕ_0 and g are regular, it is not difficult to see that Eq. (3.4) (resp. Eq. (3.5)) is equivalent to Eq. (3.2) (resp. Eq. (3.3)).

Proposition 3.1

Assume (2.33) and (2.36) and moreover that $\phi_0, g \in C(H;n) \cap K^+$, $n \in \mathbb{N}$. Then Eq. (3.4) (resp. (3.5)) has a unique solution ϕ^α (resp. $\phi^{\alpha,i}$) $\subset C([0,T];C(H;n))$ $\forall T > 0$ and we have

$$|\phi^\alpha(t,\cdot)|_{0,n} \leqslant e^{t\omega_n}|\phi_0|_{0,n} + \int_0^t e^{s\omega_n}|g|_{0,n}\, ds \tag{3.6}$$

$$|\phi^{\alpha,i}(t,\cdot)|_{0,n} \leqslant e^{t\omega_n}|\phi_0|_{0,n} + \int_0^t e^{s\omega_n}|g|_{0,n}\, ds \quad . \tag{3.7}$$

Moreover, if $\overline{\phi}_0 \leqslant \phi_0$, $\overline{g} \leqslant g$ with $\overline{\phi}_0, \overline{g} \in C(H;n)$ we have

$$\overline{\phi}^\alpha(t,x) \leqslant \phi^\alpha(t,x) \tag{3.8}$$

where $\overline{\phi}^\alpha$ is the solution of Eq. (3.4) corresponding to $\overline{\phi}_0, \overline{g}$. Finally for any $x \in H$ we have

$$\lim_{i\to\infty} \phi^{\alpha,i}(t,x) = \phi^\alpha(t,x) \tag{3.9}$$

uniformly in t on the bounded sets of $[0,+\infty[$.

Proof

We proceed by successive approximations, setting

$$\phi_{(0)}(t,x) = e^{-t/\alpha}\, E\phi_0(\zeta(t,0,x)) + \int_0^t e^{-(t-s)/\alpha}\, Eg(\zeta(t,s,x))ds \tag{3.10}$$

$$\phi_{(n+1)}(t,x) = \phi_{(0)}(t,x) + \int_0^t e^{-(t-s)/\alpha}\,\frac{1}{\alpha}\,E(\phi_{(n)})_\alpha(\zeta(t,s,x))\,ds \tag{3.11}$$

then, using (2.21) and (2.38) we can prove, by recurrence, the estimate:

$$|\phi_{(m)}(t,\cdot)|_{0,n} \leqslant e^{\omega_n t}|\phi_0|_{0,n} + \int_0^t e^{\omega_n s}|g|_{0,n}ds \overset{\text{def}}{=} N .$$

Moreover, by (2.24) and (2.38) it follows that

$$|\phi_{(m+1)}(t,\cdot) - \phi_{(m)}(t,\cdot)|_{0,n} \leqslant \int_0^t e^{-(t-s)/\alpha + \omega_n(t-s)} \cdot$$

$$\cdot \frac{1}{\alpha}(1+\alpha C_n(N))|\phi^{(m)}(s,\cdot) - \phi^{(m-1)}(s,\cdot)|_{0,n}\,ds .$$

Now, by standard arguments, it follows that $\{\phi_{(n)}\}$ converges in $C(H;n)$ to a solution ϕ^α of Eq. (3.4). The estimate (3.6) is a consequence of (2.21) and of the Gronwall lemma. Moreover it is easy to prove that

$$\overline{\phi}_{(n)}(t,x) \leqslant \phi_{(n)}(t,x)$$

which implies (3.8). Finally (3.9) follows from Lemma 2.11 #

By using (2.22), (2.23), (2.24) we can obtain additional estimates for ϕ^α .

<u>Proposition 3.2</u>

<u>Assume the hypotheses of Proposition</u> 3.1. <u>Then the following holds:</u>

a) <u>If</u> $\phi_0, g \in C^1(H;n,m)$ $n,m \in \mathbb{N}$, $n > m$, <u>we have</u>

$$|\phi^\alpha(t,\cdot)|_{1,m} \leqslant e^{\omega_n t}|\phi_0|_{1,m} + \int_0^t e^{\omega_n s}|g|_{1,m}\,ds \tag{3.12}$$

b) <u>For</u> $m,p \in \mathbb{N}$, $L > 0$ <u>there exists</u> $\gamma = \gamma_{n,p,L}$ <u>such that if</u> $\phi_0, g \in C^1_{\mathrm{Lip}}(H;n,m,p)$ <u>and</u> $|\phi_0|_{1,n} \leqslant L, |g|_{1,m} \leqslant L$ <u>we have</u>

$$\|\phi^\alpha(t,\cdot)\|_{1,p} \leqslant e^{\gamma t}\|\phi_0\|_{1,p} + \int_0^t e^{\gamma s}\|g\|_{1,p}\,ds \tag{3.13}$$

c) <u>For</u> $L > 0$, <u>there exists</u> $\eta = \eta_{n,L}$ <u>such that if</u> $|\phi_0|_{0,n}, |\overline{\phi}_0|_{0,n}, |g|_{0,n}, |\overline{g}|_{0,n} \leqslant L$, <u>we have</u>

$$|\phi^{\alpha}(t,\cdot) - \overline{\phi}^{\alpha}(t,\cdot)|_{0,n} \leq e^{\eta t}|\phi_0 - \overline{\phi}_0|_{0,n} + \int_0^t e^{\eta s}|g - \overline{g}|_{0,n}ds. \tag{3.14}$$

For any $\phi_0, g \in C(H;n)$ we shall set in the sequel:

$$\Lambda^{\alpha}(\phi_0,g) = \phi^{\alpha} \tag{3.15}$$

where ϕ^{α} is the solution of Eq. (3.4).

We will study the limit of Λ^{α} as α goes to 0.

Proposition 3.3

Assume (2.33) and (2.36). Let $\phi_0, g \in C^1_{Lip}(H;n,m,p)$ with $n \geq 2m + p$; then there exists the limit:

$$\lim_{\alpha \to 0} \Lambda^{\alpha}(\phi_0,g) \overset{\text{def}}{=} \Lambda(\phi_0,g) \tag{3.16}$$

in $C([0,T];C(H;n))$. Moreover

$$\Lambda(\phi_0,g) \in C([0,T];C^1_{Lip}(H;n,m,p)) \tag{3.17}$$

Proof

Let $\alpha,\beta > 0$, then we can prove (by approximating ϕ^{β} with $\phi^{\beta,i}$) the equality

$$\phi^{\beta}(t,x) = e^{-t/\alpha} E\phi_0(\zeta(t,0,x)) + \\ + \int_0^t e^{-(t-s)/\alpha} E[\frac{1}{\alpha}\phi^{\beta}_{\alpha} + g + R_{\phi^{\beta},\alpha} - R_{\phi^{\beta},\beta}](\zeta(t,s,x))ds$$

Let $N = (e^{\omega_n T}|\phi_0|_{0,n} + \int_0^T e^{\omega_n s}|g|_{0,n}ds)$; then by (2.24) and (2.31) we have

$$|\phi^{\alpha}(t,\cdot) - \phi^{\beta}(t,\cdot)|_{0,n} \leq \int_0^t e^{-(t-s)/\alpha+\omega_n(t-s)}[\frac{1}{\alpha}|\phi^{\alpha}(s,\cdot) - \phi^{\beta}(s,\cdot)|_{0,n} \cdot \\ \cdot (1 + \alpha C_n(N)) + (\alpha + \beta)D(L)|\phi^{\beta}(s,\cdot)|^2_{1,m}\|\phi^{\beta}(s,\cdot)\|_{1,p}]ds.$$

On the other hand, by (3.12) and (3.13) there exists $\nu > 0$ such that

$$D(L)|\phi^{\beta}(s,\cdot)|^2_{1,m}\|\phi^{\beta}(s,\cdot)\|_{1,p} \leq \nu$$

so that, by the Gronwall lemma, we get:

$$|\phi^{\alpha}(t,\cdot) - \phi^{\beta}(t,\cdot)|_{0,n} \leq \nu\, e^{\omega + C_n(N)} (\alpha + \beta)$$

thus the limit (3.16) does exist. Finally (3.17) follows from an Ascoli-Arzelà argument (see [1]) #

Let now $n \geq 2$. We notice that, by Lemma 2.8, $C^1_{Lip}(H;2,1,1) \cap K^+$ is dense in $C(H;n) \cap K^+$; thus by Proposition 3.3 we can extend, by density, Λ to a mapping:

$$\Lambda : [C(H;n) \cap K^+] \times [C(H;n) \cap K^+] \to C([0,T] \times H;n) . \tag{3.19}$$

By Propositions 3.1, 3.2, 3.3 we get now the following

Proposition 3.4

Λ has the following properties:

a) For any $n \geq 2$, ϕ_0, $g \in C(H;n) \cap K^+$ we have:

$$|\Lambda(\phi_0,g)(t)|_{0,n} \leq e^{\omega_n t} |\phi_0|_{0,n} + \int_0^t e^{\omega_n s} |g|_{0,n} ds \tag{3.20}$$

b) For any $n \geq 2$, $L > 0$, there exists $\eta = \eta_{n,L}$ such that if $|\phi_0|_{0,n}$, $|\overline{\phi}_0|_{0,n}$, $|g|_{0,n}$, $|\overline{g}|_{0,n} \leq L$, we have

$$|\Lambda(\phi_0,g)(t) - \Lambda(\overline{\phi}_0,\overline{g})(t)|_{0,n} \leq e^{\eta t}|\phi_0 - \overline{\phi}_0|_{0,n} + \int_0^t e^{\eta s}|g - \overline{g}|_{0,n} ds \tag{3.21}$$

c) If ϕ_0, $\overline{\phi}_0$, g, $\overline{g} \in C(H;n) \cap K^+$ and $\phi_0 \leq \overline{\phi}_0$, $g \leq \overline{g}$ we have $\Lambda(\phi_0,g) \leq \Lambda(\overline{\phi}_0,\overline{g})$

d) If ϕ_0, $g \in C^1_{Lip}(H;n,m,p)$ with $n \geq 2m + p$ then $\Lambda(\phi_0,g) \in C([0,T];C^1_{Lip}(H;n,m,p))$.

The following result will be useful to study Dynamic Programming.

Proposition 3.5

Assume that ϕ_0, $g \in C^1_{Lip}(H;n,m,p) \cap K^+$ with $n \geq 2m + p$. Let $\phi = \Lambda(\phi_0,g)$, $u \in M^2_W(0,T,H)$ and let y be the solution of (1.4). Then we have

$$\phi(t,x) + E \int_0^t |u(s) + \phi_x(s,y(s))|^2 ds =$$

$$= E \left\{ \int_0^t [g(y(s)) + \frac{1}{2} |u(s)|^2] ds + \phi_0(y(T)) \right\} . \tag{3.22}$$

Moreover

$$\phi(t,x) = E\left\{\int_0^t [\,g(y^\star(s)) + \frac{1}{2}\,|u^\star(s)|^2] + \phi_0(y^\star(T))\right\} \tag{3.23}$$

where $y^\star$ is the solution of the equation

$$dy = (Ay - \phi_x(t,y))dt + \sqrt{\varepsilon}\; By\, dw_t \quad , \quad y(0) = x \tag{3.24}$$

and $u^\star$ is given by

$$u^\star(s) = -\phi_x(s,y^\star(s)) \quad .$$

Proof

It is not difficult to prove that

$$\phi^{\alpha,i}(t,x) + E\int_0^t |P_i u(s) + \phi_x^{\alpha,i}(s,y_i(s))|^2 ds =$$

$$= E\left\{\int_0^t [g(y_i(s)) + \frac{1}{2}\,|P_i u(s)|^2 + |R_{\phi^{\alpha,i},\alpha}(y_i(s))|]ds - \phi_0(y_i(t))\right\}$$

where $\phi^{\alpha,i}$ is the solution of Eq. (3.5) and y_i is the solution of

$$dy_i = (A_i y_i + P_i u)dt + \sqrt{\varepsilon}\; B_i y_i\, dw_t^i \quad , \quad y_i(0) = P_i x \;.$$

Now equality (3.22) follows from (2.31), (3.9) and (3.16). The other statements can be proved as in [2] #

Remark 3.6

In [1] it is proved that if ϕ_0, $g \in C^2_{Lip}(H;n,m,p,q)$ with $n \geqslant 2m(1+p)$ then $\phi = \Lambda(\phi_0,g)$ is a classical solution of Eq. (1.1).

4. BELLMAN STATIONARY EQUATION

We assume here (2.33) and (2.36) and consider the equation:

$$\frac{1}{2}\,|\phi_x|^2 - \frac{\varepsilon}{2}\,\mathrm{Tr}(\phi_{xx}(Bx)S(Bx)^\star) - \langle Ax,\phi_x\rangle = g \;. \tag{4.1}$$

We make the following assumptions on g :

$$\begin{cases} \text{i)} \quad g \in C^1_{Lip}(H;n,m,p) \cap K^+ \quad \underline{\text{with}} \quad n \geq 2m+p \\ \text{ii)} \quad \underline{\text{There exist}} \quad C_1, C_2 \quad \underline{\text{such that}} \quad C_1|x|^2 \leq g(x) \leq C_2(|x|^2+|x|^n) . \end{cases} \tag{4.2}$$

Our purpose is to find a solution of Eq. (4.1) taking the limit as $t \to \infty$ of $\phi(t,x) = \Lambda(0,g)(t,x)$. Remark that $\phi(t,x)$ is increasing in t for any $x \in H$. In fact let $\varepsilon > 0$, $S(t,x) = \phi(t+\varepsilon,x)$, that is $S = \Lambda(\phi(\varepsilon,\cdot),g)$; thus, by Proposition 3.4-c)

$$S(t,x) \geq \Lambda(0,g)(t,x) = \phi(t,x) .$$

We set:

$$\phi_\infty(x) = \lim_{t\to\infty} \phi(t,x) \qquad \forall x \in H \tag{4.3}$$

<u>Proposition 4.1</u>

ϕ_∞ <u>belongs to</u> $C(H;n)$. <u>Moreover for any</u> $x \in H$ <u>there exist</u> $t_n \uparrow \infty$ <u>and</u> $z \in \partial\phi_\infty(x)$ <u>such that</u>

$$\phi_x(t_n,x) \longrightarrow z \quad \underline{\text{weak in}}\ H . \tag{4.4}$$

<u>Proof</u>

Fix $\lambda > 0$ and set

$$u(t) = -\lambda e^{-\lambda t}\zeta(t,0,x) , \qquad y(t) = e^{-\lambda t}\zeta(t,0,x) \tag{4.5}$$

where ζ is defined by (2.37).

Using Ito's formula it is easy to find $\eta \in \mathbb{R}$ and $\bar{c} > 0$ such that

$$\begin{cases} E|u(s)|^2 \leq \lambda^2\bar{c}\, e^{-(\lambda-\eta)s}\, |x|^2 \\ Eg(y(s)) \leq \bar{c}\, e^{-(\lambda-\eta)s}(|x|^2+|x|^n) . \end{cases} \tag{4.6}$$

From (3.22) it follows that there exists $h \in C(H;n) \cap K^+$ such that $\phi(t,x) \leq h(x)$ and all the conclusions follow from standard arguments #

<u>Proposition 4.2</u>

<u>We have</u>

$$\Lambda(\phi_\infty, g) = \phi_\infty \quad . \tag{4.7}$$

<u>Proof</u>

Set

$$\phi_k(t,x) = \phi(t+k,x) = \Lambda(\phi(k,\cdot),g) = \Lambda(\phi_{0k},g) \tag{4.8}$$

$$\overline{\phi} = \Lambda(\phi_\infty, g) \tag{4.9}$$

then:

$$\lim_{k\to\infty} \phi_{0k}(x) = \phi_\infty(x) \tag{4.10}$$

and clearly

$$\phi_k(t,x) \uparrow \phi_\infty(x) \leqslant \overline{\phi}(t,x) \tag{4.11}$$

thus we have only to prove that $\phi_\infty = \overline{\phi}$.

Now, by (3.23) we have

$$\phi_k(t,x) = E\left\{\int_0^t [\, g(y_k(s)) + \frac{1}{2}\, |u_k(s))|^2 \,]\, ds + \phi_{0k}(y_k(t))\right\} \tag{4.12}$$

where $u_k = -\phi_{kx}(t,y_k)$ and

$$\begin{cases} dy_k = (Ay_k - u_k)dt + \sqrt{\varepsilon}\, By_k\, dw_t \\ y_k(0) = x \quad . \end{cases}$$

By using Ito's formula it is not difficult to prove that

$$E\,|y_k(t)|^2 \leqslant C_T \quad , \quad E\,|u_k(t)|^2 \leqslant C_T$$

where C_T is a suitable constant. Thus there exist $\tilde{y}$ and $\tilde{u}$ such that $y_k \longrightarrow \tilde{y}$, $u_k \longrightarrow \tilde{u}$ weakly in $M_w^2(0,T;H)$. By (4.12) it follows

$$\phi_\infty(x) \geqslant E\left\{\int_0^t [\, g(\tilde{y}) + \frac{1}{2}\, |\tilde{u}|^2 \,]\, ds + \phi_\infty(y(t)) \geqslant \overline{\phi}(t,x)\right\} \quad \#$$

A solution ϕ_∞ to Eq. (4.7) can be viewed as a weak solution to Eq. (4.1).

We will show now that it is possible to perform Dynamic Programming, for the infinite horizon problem (1.3). For this we need first two lemmas:

Lemma 4.3

For any $T > 0$ *there exists a unique solution* y_T *of the problem:*

$$\begin{cases} dy_T = (Ay_T - \phi_x(T-t, y_T))dt + \sqrt{\varepsilon}\, By_T\, dw_t \\ y_T(0) = x \quad , \quad t \in [0,T] \ . \end{cases} \tag{4.13}$$

Moreover there exists $c' > 0$ *such that*

$$E \int_0^T |y_T(t)|^2 dt \leq c' \phi_\infty(x) \tag{4.14}$$

$$E \int_0^T |\phi_x(T-t, y_T)|^2 dt \leq c' \phi_\infty(x) \quad . \tag{4.15}$$

Proof

The existence follows from a result of Tubaro ([11]). Moreover, setting $u = u_T$ in (3.22) we get

$$\begin{aligned} \phi(t,x) &= E \int_0^t [g(y_T(s)) + \tfrac{1}{2} |u_T(s)|^2] ds \geq \\ &\geq E \int_0^t [c_1 |y_T(s)|^2 + \tfrac{1}{2} |u_T(s)|^2] ds \end{aligned} \tag{4.16}$$

which implies (4.14) and (4.15) #

Lemma 4.4

There exist y^* *and* u^* *solutions of the equation*

$$\begin{cases} dy^* = (Ay^* + u^*)dt + \sqrt{\varepsilon}\, By^*\, dw_t & t > 0 \\ y^*(0) = x \end{cases} \tag{4.17}$$

and such that

$$u^*(t) \in -\partial\phi_x^\infty(y^*(t)) \qquad \forall t \geq 0 \quad . \tag{4.18}$$

Moreover there exists $c > 0$ *such that*

$$E\left(\int_0^\infty |y^*(t)|^2 dt + \int_0^\infty |u^*(t)|^2 dt \right) \leq c\phi_\infty(x) \quad . \tag{4.19}$$

Finally for any $T_1 > 0$ *we have*

$$y_T \longrightarrow y^\star \quad , \quad - \phi_x(T - t, y_T) \longrightarrow u^\star \tag{4.20}$$

<u>as</u> $T \to \infty$ <u>weakly in</u> $M^2_W(0,T_1;H)$.

<u>Proof</u>

Fix $T_1 > 0$. By (4.15) and (4.16) there exist $y^\star$ and $u^\star$ such that (4.20) holds. Moreover we have

$$y_T(x) = \zeta(t,0,x) - \int_0^t \zeta(t - s, \phi_x(T - t, y_T(s))ds \tag{4.21}$$

as $T \to \infty$ we find

$$y^\star(t) = \zeta(t,0,x) - \int_0^t \zeta(t - s, u^\star(s))ds$$

and (4.17) follows #

We can now prove

<u>Proposition 4.5</u>

<u>Assume</u> (2.33), (2.36) <u>and</u> (4.2) <u>and let</u> ϕ^∞ <u>be defined by</u> (4.3). <u>Then</u> ϕ_∞ <u>coincides with the value function</u> J_∞ :

$$J_\infty(x) = \inf\{J(x,u) \ ; \ u \in M^2_W(0,\infty;H)\} \quad . \tag{4.22}$$

<u>Moreover there exists a unique optimal control</u> $u^\star$ <u>related to the optimal state</u> $y^\star$ <u>by the synthesis formula:</u>

$$u^\star(t) \in - \partial\phi^\infty_x(y^\star(t)) \qquad t \geq 0 \quad . \tag{4.23}$$

<u>Proof</u>

By (3.22) we have

$$\phi(t,x) \leq E\int_0^t [g(y(s)) + \frac{1}{2}|u(s)|^2]ds \qquad \forall u \in M^2_W(0,\infty;H)$$

so that $\phi(x) \leq J_\infty(x)$.

Conversely setting in (3.22) $y = y_T$, $u = u_T$, where y_T is the solution of (4.13) and $u_T = - \phi_x(T - t, y_T)$, we get

$$\phi(T,x) = E\int_0^t [g(y_T(s)) + \frac{1}{2}|u_T(s))|^2]ds \ .$$

Fix $\overline{T}_1 > 0$, choose $T \geq \overline{T}_1$ and let $T \uparrow \infty$, then

$$\phi(T,x) \geq E \int_0^{\bar{T}_1} [g(y^\star(s)) + \frac{1}{2}|u^\star(s)|^2]\,ds$$

which implies $\phi_\infty(x) \geq J_\infty(x)$ so that $\phi_\infty = J_\infty$.

We prove now (4.22) and uniqueness of optimal control. Assume that $(\tilde{u},\tilde{y})$ is an optimal couple, then, again by (3.22) we have

$$\phi(T,x) + \frac{1}{2} E \int_0^T |\tilde{u}(s) + \phi_x(T-s,\tilde{y}(s))|^2\,ds =$$

$$= \phi_\infty(x) - \int_T^\infty [g(\tilde{y}(s)) + \frac{1}{2}|\tilde{u}(s)|^2]\,ds \tag{4.24}$$

as $T \to \infty$ we have:

$$\lim_{T\to\infty} \int_0^T |\tilde{u}(s) + \phi_x(T-s,\tilde{y}(s))|^2\,ds = 0$$

so that, for any $T_1 > 0$, $\phi_x(T-s,\tilde{y}(s)) \to -\tilde{u}(s)$ in $M_W^2(0,T_1;H)$. Thus $\tilde{u}(s) \in -\partial\phi_x^\infty(s,\tilde{y}(s))$ and $\tilde{u}(s) = u^\star$ by virtue of uniqueness for the problem

$$\begin{cases} dy \in (Ay^\star - \partial\phi_x^\infty(y^\star))dt + \sqrt{\varepsilon}\,By^\star dw_t \\ y(0) = x \end{cases}$$

(see [11]).

REFERENCES

[1] V. BARBU-G. DA PRATO, Hamilton Jacobi Equations in Hilbert Spaces, PITMAN, London (1983).

[2] V. BARBU-G. DA PRATO, Solution of the Bellman Equation Associated with an Infinite Dimensional Stochastic Control Problem and Synthesis of Optimal Control, SIAM J. Control and Optimization, 21, 4 (1983) 531-550.

[3] V. BARBU-Th. PRECUPANU, Convexity and Optimization in Banach Spaces, SIJTHOFF and NOORDHOFF, Gröningen, (1978).

[4] G. DA PRATO, Direct Solution of the Bellman Equation for a Stochastic Control Problem, Control Theory for Distributed Parameter Systems and Applications Edited by F. KAPPEL, K. KUNISCH, W. SCHAPPACHER, SPRINGER-VERLAG (1983), Lecture Notes in Control and Information Sciences vol. 54, 92-99.

[5] W.H. FLEMING-R.W. RISHEL, Deterministic and Stochastic Optimal Control, Springer-Verlag, New York, (1975).

[6] A. ICHICAWA, Linear Stochastic Evolution Equations in Hilbert Spaces, J. DIFF. EQUAT. 28 (1978) 266-283.

[7] P. KOTELENEZ, A submartingale type inequality with applications to stochastic evolution equations, STOCHASTICS, 8 (1982) 139-151.

[8] M. METIVIER-J. PELLAUMAIL, Stochastic Integral, Academic Press (1977).

[9] E. PARDOUX, Equations aux dérivées partielles stochastiques non linéaires monotones. Thèse, Université Paris XI (1975).

[10] L. TARTAR, Sur l'étude directe d'équation non linéaires intervenant en Théorie du Contrôle Optimal, J. FUNCTIONAL ANALYSIS 17 (1974), 1-47.

[11] L. TUBARO, On Abstract Stochastic Differential Equation in Hilbert Spaces with Dissipative Drift, STOCHASTIC ANALYSIS AND APPLICATIONS, 4 (1983), 205-214.

A STOCHASTIC CONTROL APPROACH

TO SOME LARGE DEVIATIONS PROBLEMS

Wendell H. Fleming[1)]
Lefschetz Center for Dynamical Systems
Division of Applied Mathematics
Brown University
Providence, Rhode Island 02912

1. Introduction

The subject of large deviations is concerned with asymptotic formulas for exponentially small probabilities and expectations, associated with stochastic processes. Such large deviations problems are typically formulated in terms of a family x_t^ε of processes depending on a small positive parameter ε. Let $P^\varepsilon = P(A^\varepsilon)$ be the probability of some event A^ε depending on the sample path $x_\cdot^\varepsilon$. If $-\varepsilon \log P^\varepsilon$ tends to a positive limit $I^0 > 0$, then there is a large deviation. P^ε is exponentially small, roughly of order $\exp(-\varepsilon^{-1} I^0)$. Usually, the limit I^0 turns out to be characterized as the minimum in a certain associated optimization problem. More generally, if E^ε is the expectation of some functional of $x_\cdot^\varepsilon$, then one may have a large deviations result of the form $-\varepsilon \log E^\varepsilon \to I^0$ as $\varepsilon \to 0$. Again I^0 is usually characterized in terms of a minimization problem.

In general terms, our approach to such large deviations questions is as follows. Let

$$I^\varepsilon = -\varepsilon \log E^\varepsilon. \tag{1.1}$$

We seek to characterize I^ε as the minimum in a suitable stochastic control problem. To obtain a large deviations theorem one then has to solve the technical problem of showing that, as $\varepsilon \to 0$, the minimum I^ε tends to the minimum I^0 for a corresponding "limiting" control problem. Up to now this program has only been carried out in some special cases.

In Section 2 we suppose that x_t^ε is a nearly deterministic Markov diffusion process in n-dimensional R^n, stopped at the boundary ∂D of some bounded open set $D \subset R^n$. We consider large deviations problems of Ventsel-Freidlin type [14] for which E^ε can be found by solving a boundary value problem for a linear second order PDE of parabolic type

---

1) Supported by the National Science Foundation under Grant No. MCS 8121940, by the Air Force Office of Scientific Research under Grant No. AF-AFOSR 81-0116, and by the Office of Naval Research under Grant No. N00014-83-K-0542.

(the backward equation for the process x_t^ε). The logarithmic transformation (1.1) changes the backward equation into a nonlinear parabolic PDE for I^ε, which is the dynamic programming equation for a controlled diffusion process. When $\varepsilon = 0$ the dynamic programming equation becomes the Hamilton-Jacobi equation for a corresponding problem in calculus of variations. In Section 3 we specialize to large deviations for the exit probability $P^\varepsilon = P(\tau^\varepsilon \leq T)$, where τ^ε is the first time t such that $x_t^\varepsilon \in \partial D$. It is shown later (in Section 5) that the optimal control law turns out to have a natural interpretation as the drift of the process obtained by conditioning on the event $A^\varepsilon = \{\tau^\varepsilon \leq T\}$. This fact was pointed out to the author by M. Day.

In Section 4 we outline how the large deviations result $I^\varepsilon \to I^0$ as $\varepsilon \to 0$ can be proved in this setting by PDE-viscosity solution methods. The idea of proving such results by stochastic control methods goes back to [6]. The use of PDE-viscosity solution methods for such large deviations problems was initiated by Evans-Ishii [4]. The result in Section 3 is a particular case of results proved in [8].

In Sections 5-7 we turn from nearly deterministic Markov diffusions to some questions about more general classes of Markov processes x_t. In Section 5 we introduce a change of probability measure depending on a weight factor $\Phi(x_T)$ for a specified final time T. This leads to a corresponding change in backward operator for the process x_t. In Section 6 we give a stochastic control representation for expressions of the form $I = -\log E\,\Phi(x_T)$. The formulation of the stochastic control problem is that of Sheu [12]. It turns out that the optimal control leads to the same backward operator as the one obtained in Section 5 by change of probability measure.

In Section 7 we consider some problems involving functionals of the occupation measure μ_t, which we call large deviations problems of Donsker-Varadhan type. Here a single Markov process x_t is given. The small parameter T^{-1} $(T \to \infty)$ has the role of ε. The goal is to obtain results like those of Donsker-Varadhan [1] using logarithmic transformations like (1.1) and stochastic control techniques. In stochastic control terms, it suffices to show that under suitable assumptions the minimum in a certain finite time horizon control problem (7.11) tends as $T \to \infty$ to the minimum in a corresponding equilibrium control problem (7.13). However, this has as yet been done only in the special case of the Donsker-Varadhan formula for the dominant eigenvalue of $L + V(x)$ where L is the generator of x_t. See Holland [10],[11] and Sheu [13].

2. Nearly deterministic Markov diffusions.

Let x_t^ε be an n-dimensional Markov diffusion process, satisfying the stochastic differential equation on an interval $s \leq t < \infty$

$$dx_t^\varepsilon = b(x_t^\varepsilon)dt + \sqrt{\varepsilon}\sigma(x_t^\varepsilon)dw_t, \quad x_s^\varepsilon = x, \quad x \in R^n, \quad s \geq 0, \tag{2.1}$$

with w_t an n-dimensional Brownian motion. We assume that b, σ are Lipschitz continuous functions. Moreover, the matrices $a(x) = \sigma(x)\sigma'(x)$ satisfy for some $c > 0$

$$\sum_{i,j=1}^{n} a_{ij}(x)\lambda_i\lambda_j \geq c|\lambda|^2, \quad \forall \lambda \in R^n.$$

Let

$$L^\varepsilon f = \frac{\varepsilon}{2}\sum_{i,j} a_{ij}(x) f_{x_i x_j} + b(x)\cdot\nabla f. \tag{2.2}$$

Then L^ε coincides with the generator of x_t^ε for $f \in C_b^2(R^n)$, i.e., for f with bounded continuous partial derivatives of orders 0,1,2.

Let $D \subset R^n$ be bounded, open with smooth boundary ∂D. Let $\bar{x}_t^\varepsilon$ be the process stopped at ∂D, namely,

$$\bar{x}_t^\varepsilon = x_t^\varepsilon, \quad \text{if} \quad t \leq \tau^\varepsilon$$

$$= x_{\tau^\varepsilon}^\varepsilon, \quad \text{if} \quad t > \tau^\varepsilon,$$

where τ^ε $(=\tau_{sx}^\varepsilon)$ is the exit time of x_t^ε from D starting from $x \in D$ at time $s \geq 0$. Let $G \in C^2(\bar{D})$. For fixed $T > 0$ and $0 \leq s < T$, let

$$E^\varepsilon(s,x) = E_{sx}[\exp(-\frac{G(\bar{x}_T^\varepsilon)}{\varepsilon})].$$

We are interested in a large deviation result for E^ε. Let

$$Q = [0,T) \times D, \quad \partial Q = ([0,T] \times \partial D) \cup (\{T\} \times D).$$

Then $E^\varepsilon(s,x)$ is the unique solution to the linear parabolic partial differential equation (backward equation)

$$\frac{\partial E^\varepsilon}{\partial s} + L^\varepsilon E^\varepsilon = 0 \quad \text{in} \quad Q \tag{2.3}$$

with boundary conditions on ∂Q

$$E^\varepsilon(s,x) = \exp(-\frac{G(x)}{\varepsilon}). \tag{2.4}$$

Moreover, E^ε is a smooth function of (s,x) in the sense that $E^\varepsilon \in C^{1,2}(\bar{Q} - \{T\} \times \partial D)$ and $E^\varepsilon, \nabla_x E^\varepsilon$ are continuous on $\bar{Q}$. See [7, Appendix E] for example. The function $I^\varepsilon = -\varepsilon \log E^\varepsilon$ satisfies

$$\frac{\partial I^\varepsilon}{\partial s} + L^\varepsilon I^\varepsilon - \frac{1}{2}(\nabla I^\varepsilon)' a(x)(\nabla I^\varepsilon) = 0 \quad \text{in} \quad Q, \tag{2.5}$$

$$I^\varepsilon(s,x) = G(x) \quad \text{on} \quad \partial Q, \tag{2.6}$$

where for brevity we write $\nabla = \nabla_x$ for the gradient in the variables

x. For $\varepsilon = 0$, the analogue of (2.5) is the first order partial differential equation

$$\frac{\partial I^0}{\partial s} + b(x)\cdot\nabla I^0 - \frac{1}{2}(\nabla I^0)'a(x)(\nabla I^0) = 0. \tag{2.7}$$

This is the Hamilton-Jacobi equation for the following calculus of variations problem. Let

$$k(x,u) = \frac{1}{2}(u-b(x))'a^{-1}(x)(u-b(x)). \tag{2.8}$$

Consider the class of $\eta \in C^1([s,T];R^n)$ such that $\eta_s = x$. Let θ denote the exit time of η_t from D, and $\theta \wedge T = \min(\theta,T)$. Let

$$I^0(s,x) = \inf_{\eta}\{\int_s^{\theta\wedge T} k(\eta_t,\dot\eta_t)dt + G(\eta_{\theta\wedge T})\}. \tag{2.9}$$

Let us make the additional assumption

$$G(x) = 0 \quad \text{for} \quad x \in \partial D, \; G(x) > 0 \quad \text{for} \quad x \in D. \tag{2.10}$$

It then follows from the theory of viscosity solutions that

$$I^0 = \lim_{\varepsilon\to 0} I^\varepsilon, \tag{2.11}$$

and that $I^0(s,x)$ is the unique Lipschitz continuous viscosity solution of (2.8) with the boundary condition (2.6).

The PDE-viscosity solution method of proving (2.11) will be indicated in Section 4. In this argument, assumption (2.10) is used to obtain an a priori estimate for ∇I^ε for $x \in \partial D$, $0 \leq s < T$.

The PDE-viscosity solution method makes no use of stochastic control techniques. An alternative stochastic control proof of (2.11) can be given, using the fact that (2.5) is the dynamic programming equation for the following stochastic control problem. The state ξ_t of the process being controlled satisfies the stochastic differential equation

$$d\xi_t = \underline{u}(t,\xi_t)dt + \sqrt{\varepsilon}\sigma(\xi_t)dw_t, \quad s \leq t \tag{2.12}$$

$$\xi_s = x$$

where $u_t = \underline{u}(t,\xi_t)$ is the control applied at time t. The feedback control law $\underline{u}(\cdot,\cdot)$ is assumed bounded and Borel measurable. Let θ^ε be the exit time of ξ_t from D. A verification theorem in stochastic control theory [7, VI.4] implies that the solution I^ε to (2.5)-(2.6) has the following representation:

$$I^\varepsilon(s,x) = \min_{\underline{u}} E_{sx}\{\int_s^{\theta^\varepsilon\wedge T} k(\xi_t,u_t)dt + G(\xi_{\theta^\varepsilon\wedge T})\}. \tag{2.13}$$

Moreover, the feedback control

$$\underline{u}^{\varepsilon}(s,x) = b(x) - a(x)\nabla I^{\varepsilon}(s,x), \quad (s,x) \in Q, \tag{2.14}$$

is optimal. The change of drift from b in (2.1) to $\underline{u}^{\varepsilon}$ in (2.12) corresponds to a change of probability measure which will arise in Section 5.

In the stochastic control proof that $I^{\varepsilon} \to I^0$ as $\varepsilon \to 0$ condition (2.10) is not in fact needed. In [6, Lemma 7.1] it is shown that

$$I^0 \le \liminf_{\varepsilon\to 0} I^{\varepsilon}. \tag{2.15}$$

The inequality

$$\limsup_{\varepsilon\to 0} I^{\varepsilon} \le I^0$$

can be proved by the following argument, which we merely sketch. In (2.9) we may take the infimum in the class of η such that either: (i) $\theta > T$, or (ii) $\theta < T$ and $\dot{\eta}_{\theta}$ is not tangent to ∂D at η_{θ}. We use the open loop control $\underline{u}(t) = \dot{\eta}_t$ in the stochastic control problem. For small ε the corresponding ξ_t in (2.12) is near η_t uniformly on $[s,T]$, with probability nearly 1. Moreover, in case (ii) $\theta^{\varepsilon} \to \theta$ in probability as $\varepsilon \to 0$. For such η, the expectation in (2.13) tends to the corresponding quantity in (2.9) as $\varepsilon \to 0$, from which (2.16) follows.

Without assumption (2.10) one cannot expect I^0 to be a Lipschitz function on $\bar{Q}$. It may happen when (2.10) does not hold that I^0 fails to take the boundary value $G(x_1)$ as $x \to x_1, x_1 \in \partial D$.

3. Exit probabilities.

For $(s,x) \in Q$ consider the exit probability

$$P^{\varepsilon}(s,x) = P_{sx}(\tau^{\varepsilon} \le T), \tag{3.1}$$

and let

$$I^{\varepsilon} = -\varepsilon \log P^{\varepsilon}. \tag{3.2}$$

We now take $G(x) = 0$ for $x \in \partial D$, $G(x) = +\infty$ for $x \in D$. Then $P^{\varepsilon} = E^{\varepsilon}$ in Section 2. The function I^{ε} is again a smooth solution to (2.5) in $\bar{Q} - \{T\} \times \bar{D}$, but the boundary conditions are now

$$I^{\varepsilon}(s,x) = 0, \quad 0 \le s < T, \quad x \in \partial D \tag{3.3}$$

$$I^{\varepsilon}(T,x) = +\infty, \quad x \in D.$$

One has again the result $I^{\varepsilon} \to I^0$ as $\varepsilon \to 0$, where now

$$I^0(s,x) = \inf_{\eta} \int_s^{\theta} k(\eta_t, \dot{\eta}_t)dt \tag{3.4}$$

and the infimum is taken among all $\eta \in C^1([s,T];R^n)$ such that $\eta_s = x$

and the exit time satisfies $\theta \leq T$. In the stochastic control problem, it suffices to consider feedback controls $\underline{u}$ for which the exit time also satisfies $\theta^{\varepsilon} \leq T$. Otherwise the expression to be minimized in (2.13) is infinite since $G(x) = +\infty$ for $x \in D$. In this formulation we require that $\underline{u}(s,x)$ is bounded for $s \leq T^1 < T$. In particular, the optimal control $\underline{u}^{\varepsilon}$ in (2.14) becomes unbounded as $s \to T$, $x \to \partial D$. The function I^0 is a viscosity solution to (2.7), which is locally Lipschitz. In fact, I^0 is Lipschitz on $[0,T^1] \times \overline{D}$ for any $T^1 < T$.

If the solution to $\dot{x}^0 = b(x_t^0)$ starting from $x_s^0 = x$ does not exit from D by time T, then $I^0(s,x) > 0$. In this case, there is a large deviation for the exit probability, and $I^0(s,x)$ indicates the strength of it.

4. Indications of proof and possible extensions of results.

The PDE-viscosity solution method was first applied to large deviations problems for nearly deterministic Markov diffusions by Evans and Ishii [4]. A proof by such methods that $I^{\varepsilon} \to I^0$ as $\varepsilon \to 0$ for the problem in Section 2 is by now standard. For the case of exit probabilities in Section 3, some modifications are needed to account for the infinite boundary conditions in (3.3). This is done in [8]. In outline the proof proceeds as follows. First an upper bound for I^{ε} is obtained, uniformly on $[0,T^1] \times \overline{D}$, $T^1 < T$. This can be done by introducing suitable comparison functions, or by a simple probabilistic estimate. Next, an a priori bound for $|\nabla I^{\varepsilon}|$ is obtained uniformly on $[0,T^1] \times \overline{D}$. This is done first for $x \in \partial D$ by a barrier argument, and then in the interior of D by a version of the "Bernstein trick". For subsequences I^{ε} tends to a limit I uniformly on $[0,T^1] \times \overline{D}$, any $T^1 < T$, and I is a viscosity solution of (2.7) with the boundary condition (3.3). It is shown by a direct argument that I^0 is another viscosity solution of (2.7) with the same boundary condition (3.3). A uniqueness result contained in a forthcoming paper by M. Crandall, P-L. Lions, and P. E. Souganidis then implies that $I = I^0$ and hence $I^{\varepsilon} \to I^0$ as $\varepsilon \to 0$.

Reference [8] actually deals with a more general situation in which x_t^{ε} itself is a controlled diffusion, i.e. a control enters in the dynamics (2.1). The question then is to prove a large deivations result for the minimum exit probability. In that case, I^0 is the lower value of a corresponding differential game.

In [4, Section 3] large deviations for the nonexit probability (rather than exit probability) are considered. This problem is slightly simpler in that the $+\infty$ boundary condition in (3.3) then occurs for $s < T$, $x \in \partial D$ and the 0 boundary condition for $s = T$, $x \in D$.

A stronger result than convergence of I^{ε} to I would be an asymptotic expansion of the form

$$I^{\varepsilon} = I^0 + \varepsilon J_1 + \varepsilon^2 J_2 + \ldots + \varepsilon^m J_m + 0(\varepsilon^m) \tag{4.1}$$

for any $m \geq 1$. Such an expansion can be expected to hold only in regions where I^0 is a smooth function of (s,x). In a somewhat different setting, an expansion (4.1) was obtained by a rather complicated stochastic control technique in [5]. It would be interesting to obtain (4.1) using PDE-viscosity solution methods. The argument in [5, Section 6] indicates that a useful first step would be to show that $\nabla I^{\varepsilon} \to \nabla I^0$ as $\varepsilon \to 0$, uniformly on compact subsets of a region where we expect (4.1) to hold. The weaker result $\nabla I^{\varepsilon} \to \nabla I^0$ at each point (s,x) where I^0 is differentiable was recently proved by P. Souganidis. His proof uses known one sided estimates for second order derivatives $I^{\varepsilon}_{x_i x_i}$.

Finally, we note that if I^0 is of class C^1, then the classical method of characteristics for the Hamilton-Jacobi equation (2.7) gives the optimal control law

$$\underline{u}^0(s,x) = b(x) - a(x)\nabla I^0(s,x). \tag{4.2}$$

By comparison with (2.14), we see that convergence of ∇I^{ε} to ∇I^0 corresponds when I^0 is smooth to convergence of the optimal $\underline{u}^{\varepsilon}$ to $\underline{u}^0$.

5. Changes of backward operator and probability measure.

We now turn from Markov diffusions on R^n to some questions about more general classes of Markov processes. Let Σ be a Polish space (complete separable metric). We consider Markov processes x_t with state space Σ, defined for $s \leq t \leq T$ and with initial state $x_s = x$. Moreover, the sample paths $x_{\cdot}$ lie in $D_s = D([s,T];\Sigma)$, the space of cadlag functions (right continuous with left hand limits). Let P_{sx} be the probability law of $x_{\cdot}$; P_{sx} is a probability measure on $\mathcal{F}_T$, where $\mathcal{F}_t$ is the σ-algebra generated by paths in D_s up to time $t \leq T$.

Let $\{L_s\}$, $0 \leq s \leq T$, be a family of linear operators such that $\partial/\partial s + L_s$ is a backward evolution operator for the Markov process x_t in the following sense. Let $C_b(\Sigma)$ denote the space of continuous bounded functions on Σ, and $\mathcal{D} \subset C_b(\Sigma)$ a large enough class of functions that the expectations $E_{sx} f(x_t)$ for all $f \in \mathcal{D}$ determine P_{sx}.

Let $\mathscr{L}$ be the class of $\psi(\cdot,\cdot)$ such that $\psi \in C_b([0,T]\times\Sigma)$ and $\partial\psi/\partial s$, $L_s\psi(s,\cdot)$ are in $C_b([0,T']\times\Sigma)$ for any $T' < T$. We require that

$$M_\psi(t) = \psi(t,x_t) - \psi(s,x) - \int_s^t \left(\frac{\partial\psi}{\partial r} + L_r\psi\right)dr \tag{5.1}$$

is a $(\mathscr{F}_t, P_{sx})$ martingale for $s < t < T$. Fix $t \in (s,T]$. If $\psi \in \mathscr{L}$ satisfies the backward evolution equation

$$\frac{\partial\psi}{\partial s} + L_s\psi = 0, \quad s < t < T, \tag{5.2}$$

$$\psi(t,x) = f(x),$$

then (5.1) implies

$$\psi(s,x) = E_{sx}f(x_t). \tag{5.3}$$

Conversely, if ψ defined by (5.3) is in $\mathscr{L}$, then ψ satisfies (5.2).

Suppose that ϕ,ψ are two solutions of (5.2) with $\varphi > 0$. The quotient $v = \phi^{-1}\psi$ satisfies another backward evolution equation

$$\frac{\partial v}{\partial s} + \tilde{L}_s v = 0, \text{ where} \tag{5.4}$$

$$\tilde{L}_s v = \frac{1}{\phi}[L_s(v\phi) - vL_s\phi]. \tag{5.5}$$

This change of backward evolution operator arises from the following change of probability measure. Given $T > 0$, $\Phi \geq 0$, change the probability measure P_{sx} to $\tilde{P}_{sx}$ such that

$$\tilde{E}_{sx}f(x_t) = \frac{E_{sx}[f(x_t)\Phi(x_T)]}{E_{sx}\Phi(x_T)}, \quad s < t \leq T, \quad f \in \mathscr{D}. \tag{5.6}$$

Denote the numerator and denominator of the right side of (5.6) by $\psi(s,x)$, $\phi(s,x)$ respectively. Assume that $\psi,\phi \in \mathscr{L}$ and that $v = \psi^{-1}\psi$ is in the corresponding class $\tilde{\mathscr{L}}$. Then ϕ satisfies (5.2) for $s \leq T$. The Markov property implies

$$\psi(s,x) = E_{sx}[f(x_t)\phi(t,x_t)].$$

Hence ψ satisfies (5.2) for $s < t$ and

$$\psi(t,x) = f(x)\phi(t,x).$$

Thus, v satisfies (5.4) for $s < t$ and

$$v(t,x) = f(x), \quad f \in \mathscr{D} \tag{5.7}$$

From (5.4), (5.7), $\frac{\partial}{\partial s} + \tilde{L}_s$ is a backward evolution operator for the Markov process with probability law $\tilde{P}_{sx}$.

In particular, let us consider the stopped diffusion $\tilde{x}^\varepsilon_t$ in Section 2. In this case $L_s = L^\varepsilon$ given by (2.2). We take $\mathscr{D} = C^2(\bar{D})$. In the definition of $\mathscr{L}$ we add the condition that (5.2) hold for $x \in \partial D$ (it holds in particular if (5.2) is satisfied in $\bar{Q} - \{T\} \times \bar{D})$. A direct calculation gives from (5.5)

$$\tilde{L}_s v = \frac{\varepsilon}{2} \sum_{i,j} a_{ij} v_{x_i x_j} + [b + a\varepsilon\nabla(\log\phi)]\cdot\nabla v$$

corresponding to a change of drift from b to $b + a\varepsilon\nabla(\log\phi)$. In particular, if $\phi = E^\varepsilon$ in Section 2 (or $\phi = P^\varepsilon$ in Section 3), the new drift becomes $\underline{u}^\varepsilon = b - a\nabla I_x^\varepsilon$. This is the optimal drift (2.14) for the stochastic control problem (2.12)-(2.13). This connection between the change of probability measure according to (5.6) and optimal change of drift will be made as a general principle in Section 6.

If $\Phi(x) = 1$ for $x \in \partial D$, $\Phi(x) = 0$ for $x \in D$, then $\phi = P^\varepsilon$ is the exit probability. The measure $\tilde{P}_{sx}$ is obtained from P_{sx} by conditioning on the event $\tau^\varepsilon \leq T$. The event $\tau^\varepsilon \leq T$ is rare under P_{sx} (at least if $I^0(s,x) > 0$); however $\tau^\varepsilon \leq T$ is a certain event under $\tilde{P}_{sx}$. Under the optimal drift $\underline{u}^\varepsilon$, the sample paths $\xi_.$ of (2.12) are near the set of optimal paths $\eta_.$ for (3.4) with probability near 1.

6. Stochastic control problem.

A stochastic control problem connected with the change of probability measure in Section 5 was introduced in the Ph.D. thesis of S-J. Sheu [12]. Let x_t be a Markov process with generator L. To simplify the exposition, let us assume to begin with that the state space Σ is compact, and that L is a bounded operator on $C(\Sigma)$. For $g > 0$, $g \in C(\Sigma)$, let

$$\begin{aligned} L^g f &= \frac{1}{g}[L(fg) - fLg], \quad f \in C(\Sigma), \\ k^g &= L^g(\log g) - \frac{Lg}{g}. \end{aligned} \tag{6.1}$$

An admissible control consists of choosing $g_t \in C(\Sigma)$ for $s \leq t \leq T$, depending continuously on t. We denote such a control by $\gamma \in C([s,T];C(\Sigma))$. The backward operator for the corresponding controlled Markov process is $\frac{\partial}{\partial s} + L^{g_s}$. We denote expectations with respect to the probability law of the controlled process by E^γ. The problem is to choose a control γ which minimizes

$$J(s,x,\gamma) = E_{sx}^\gamma\{\int_s^T k^{g_t}(x_t)dt + G(x_T)\}, \tag{6.2}$$

where $G \in C(\Sigma)$ is given. This problem has the following solution. Let $\Phi = \exp(-G)$. Then

$$\phi(s,x) = E_{sx}\Phi(x_T) \tag{6.3}$$

solves the backward evolution equation

$$\frac{\partial \phi}{\partial s} + L\phi = 0, \quad 0 \le s \le T, \tag{6.4}$$

$$\phi(T,x) = \Phi(x). \tag{6.5}$$

Theorem. An optimal control γ is obtained if

$$g_s(x) = \phi(s,x). \tag{6.6}$$

The key lemma [12] needed to prove this is the following. For $I \in C(\Sigma)$ let

$$H(I) = -e^{-I}L(e^{-I}) \tag{6.7}$$

Then

$$\min_{g>0}[L^g I + k^g] = H(I) \tag{6.8}$$

and equality holds for $g = \exp(-I)$. Now take

$$I(s,x) = -\log \phi(s,x). \tag{6.9}$$

From (6.4), (6.5)

$$\frac{\partial I}{\partial s} + H(I) = 0, \quad 0 \le s \le T, \tag{6.10}$$

$$I(T,x) = G(x). \tag{6.11}$$

It then follows, by the standard proof of the Verification Theorem in stochastic control theory [7, p. 159], that

$$I(s,x) \le J(s,x,\gamma)$$

with equality when γ is defined by (6.6).

When (6.6) holds, L^{g_s} is the same as the operator L_s in (5.5). Hence we have:

Corollary. The optimal backward evolution operator is the same one obtained by change of probability measure according to (5.6).

If Σ is not compact or the generator L is unbounded, additional restrictions are needed, for instance to insure that the operators L^g and the cost function k^g are well defined. We shall not go into these matters here. However, let us return to the special case of stopped diffusions considered in Sections 2-4. We now take $f,g \in C^2(\bar{D})$, $g > 0$. A direct calculation gives (for fixed $\varepsilon > 0$)

$$L^g f = \frac{\varepsilon}{2} \sum_{i,j} a_{ij}(x) f_{x_i x_j} + \underline{u}(x)\cdot\nabla f(x), \text{ where} \tag{6.12}$$

$$\underline{u} = b + a\varepsilon\nabla(\log g), \tag{6.13}$$

$$k^g(x) = \varepsilon^{-1}k(x,\underline{u}(x)) \tag{6.14}$$

with $k(x,u)$ as in (2.8). For stopped diffusions, T should be replaced by $\theta^\varepsilon \wedge T$ in (6.2), (6.3), where θ^ε is the exit time. Formulas (6.12) -(6.14) make the connection between the stochastic control problem in Section 2 and the present one. To a control γ corresponds the feed-back control $\underline{u}(\cdot,\cdot)$ such that $\underline{u}(s,\cdot)$ satisfies (6.13) with g replaced by g_s. The controlled process was denoted by ξ_t in Section 2 but by x_t here. This notational change reflects the stochastic differential equations vs change of probability measure viewpoints.

As another example, consider a continuous time Markov chain with a finite number of states $x = 1,\ldots,N$. The generator L has the form

$$Lf(x) = \sum_{y=1}^{N} q_{xy}[f(y)-f(x)]. \tag{6.15}$$

In this case

$$L^g f(x) = \sum_{y=1}^{N} \frac{q_{xy}g(y)}{g(x)} [f(y)-f(x)] \tag{6.16}$$

$$k^g(x) = \sum_{x\neq y} [\frac{g(y)}{g(x)} \log \frac{g(y)}{g(x)} - \frac{g(y)}{g(x)} + 1]q_{xy}. \tag{6.17}$$

A control γ changes the jumping rates from q_{xy} to time-dependent rates $q_{xy}g^s(y)[g^s(x)]^{-1}$.

7. Large deviations problems of Donsker-Varadhan type.

Let x_t be Markov on the time interval $0 \leq t < \infty$, with compact state space Σ. Let $\mathscr{B}(\Sigma)$ denote the σ-algebra of Borel sets, $\mathscr{M} = \mathscr{M}(\Sigma)$ the space of finite, nonnegative measures on $\mathscr{B}(\Sigma)$, and $\mathscr{M}_1$ the space of probability measures on $\mathscr{B}(\Sigma)$. We use the notation

$$\nu(f) = \int_\Sigma f(x)d\nu(x).$$

For $t > 0$, the occupation measure μ_t is characterized by

$$\mu_t(f) = \frac{1}{t}\int_0^t f(x_s)ds \tag{7.1}$$

for all bounded, Borel measurable f. By taking $f = 1_B$, the indicator function of $B \in \mathscr{B}(\Sigma)$, we see that $\mu_t(B)$ is the fraction of time during $[0,t]$ which the process x_t spends in B. A number of interesting applications in physics lead to large deviations problems for the occupation measure process, of the following type. Given a function Ψ on $\mathscr{M}_1$ let

$$I_T = -\frac{1}{T} \log E_{0x}\{\exp(-T\,\Psi(\mu_T))\}. \tag{7.2}$$

The problem is to show that I_T tends to a limit as $T \to \infty$, and to characterize the limit. Let L be the generator of x_t, with domain $\mathscr{D}$ dense in $C(\Sigma)$. Following Donsker-Varadhan [1] let

$$\mathcal{I}(\mu) = -\inf_{\substack{f>0\\ f\in\mathcal{D}}} \mu\left(\frac{Lf}{f}\right), \tag{7.3}$$

$$I_\infty = \inf_{\mu\in\mathcal{M}_1} [\mathcal{I}(\mu) + \Psi(\mu)] \tag{7.4}$$

In [1, Theorem 4] it is proved that, for Ψ weakly continuous on $\mathcal{M}_1$,

$$I_\infty = \lim_{T\to\infty} I_T, \tag{7.5}$$

provided some additional assumptions hold. These assumptions include a hypothesis that the transition probabilities for x_t have densities with respect to some reference measure, which are bounded above and away from 0 for each $t > 0$. References [2] and [3] deal with similar results when Ψ is strongly continuous, or when Σ is not compact.

In this section we sketch some preliminary ideas about how stochastic control methods might be used to deal with this problem. At the end we state the main technical problem which needs to be solved if this approach is to work. For the special case $\Psi(\mu) = \mu(V)$, for given $V \in C(\Sigma)$, formula (7.5) implies a formula for the dominant eigenvalue of the operator $L + V$. This special case was treated by Sheu [13] using methods like those outlined here. Earlier, Holland [10], [11] obtained stochastic control representations of the dominant eigenvalue and eigenfunction of $L + V$, when L generates a nondegenerate diffusion on R^n.

As an initial step, let us seek to describe the expectation $E_{0x}\{\exp(-T\Psi(\mu_T)\}$ in terms of the solution of a backward evolution equation. The occupation measure process μ_t is not Markov, but the pair (x_t,μ_t) is Markov. It is convenient to consider any initial time $s \geq 0$ (not just $s = 0$). Instead of μ_t we consider the $\mathcal{M}$-valued process ν_t defined as follows:

$$\nu_t(f) = \nu(f) + \int_s^t f(x_r)dr \quad (\nu_s = \nu) \tag{7.6}$$

for all bounded Borel measurable f. If $s = 0$, $\nu_0 = 0$, then $\nu_t = t\mu_t$. For the moment let us proceed formally. We consider the Markov process (x_t,ν_t), with state space $\Sigma\times\mathcal{M}$. The generator $\mathcal{L}$ is described for "sufficiently smooth" functions $F(x,\nu)$ by

$$\mathcal{L}F = L_xF + D_\nu F\cdot\delta_x. \tag{7.7}$$

Here L_x is the generator of x_t, acting on $F(\cdot,\nu)$,

$$D_\nu F(x,\nu)\cdot\alpha = \frac{d}{d\lambda} F(x,\nu+\lambda\alpha)\Big|_{\lambda=0},$$

and δ_x is a Dirac measure $(\delta_x(f) = f(x).)$. Given $T > 0$ and $\Phi(x,\nu)$

we define (in analogy with (6.3))

$$\phi(s,x,\nu) = E_{sx\nu}\Phi(x_T,\nu_T). \tag{7.8}$$

We need that, for a sufficiently large class of Φ, the backward evolution equation holds:

$$\frac{\partial\phi}{\partial s} + \mathscr{L}\phi = 0, \quad s < T \tag{7.9}$$

$$\phi(T,x,\nu) = \Phi(x,\nu).$$

We then make the logarithmic transformation, for $\Phi > 0$,

$$I_T(s,x,\nu) = -\frac{1}{T}\log\phi(s,x,\nu).$$

From (7.9) one gets the following nonlinear equation for I:

$$\frac{\partial I_T}{\partial s} + \frac{1}{T}H_x(TI_T) + D_\nu I_T\cdot\delta_x = 0, \tag{7.10}$$

where H_x is as in (6.7) with $L = L_x$. The term $D_\nu(\ )\cdot\delta_x$ in (7.7) represents a deterministic action in the variable ν, and is not affected by the logarithmic transformation. As in Section 6, a control γ assigns for $s \le t \le T$ a function $g_t(\cdot)$ on Σ. At least formally we then get, by the same method as in Section 6,

$$I_T(s,x,\nu) = \inf_\gamma E^\gamma_{sx\nu}\{\frac{1}{T}\int_s^T k^{g_t}(x_t)dt + G(x_T,\nu_T)\},$$

with k^g as in (6.1) and $G = -T^{-1}\log\Phi$. If we take in particular $G(\nu) = \Psi(T^{-1}\nu)$, $s = 0$, $\nu = 0$, then $\mu_T = T^{-1}\nu_T$ and $I_T = I_T(0,x,0)$ in (7.2). We then have the following stochastic control representation for I_T:

$$I_T = \inf_\gamma E^\gamma_{0x}\{\frac{1}{T}\int_0^T k^{g_t}(x_t)dt + \Psi(\mu_T)\}. \tag{7.11}$$

Up to now the derivation of (7.11) has been only formal. If x_t is a continuous time, finite state Markov chain, then there is no difficulty in making these calculations rigorous. The measure ν is simply a finite dimensional vector, with nonnegative components. If we take Ψ of class C^1 in ν, then (7.9) becomes a system of linear ordinary differential equations with data at time T depending smoothly on ν. The "running cost" function k^g in (7.11) is given by (6.17), for the case of a Markov chain. Formula (7.11) also holds for continuous Ψ, by approximating Ψ uniformly by functions of class C^1. We shall deal elsewhere with justifying (7.11) for other Markov processes. In doing so, one can first approximate a weakly continuous $\Psi(\mu)$ by functions of the form

$$\hat{\Psi}(\mu) = \Gamma[\mu(f_1),\dots,\mu(f_k)]$$

where Γ is a smooth function on R^K [9]. One must then verify "smoothness" as a function of (s,x,ν) of the expectation in (7.8), with

$$\Phi = \exp(-T\hat{\Gamma}(T^{-1}\nu))$$

in order to obtain rigorously (7.9), (7.10).

Now consider the following steady state analogue of the finite time minimum problem (7.11). For simplicity assume that the generator L is bounded. Let $\mathcal{G}$ be the class of $g > 0$ such that under the generator L^g the Markov process x_t has a unique equilibrium measure μ^g, with

$$\mu^g = \lim_{t\to\infty} \mu_t.$$

Let

$$J_{eq}(g) = \mu^g(k^g) + \Psi(\mu^g), \tag{7.12}$$

$$I_{eq} = \inf_{\mathcal{G}} J_{eq}(g). \tag{7.13}$$

When $g_t \equiv g$, with $g \in \mathcal{G}$, the right side of (7.12) is the limit as $T \to \infty$ of the expression in braces in (7.11). We then have

$$\limsup_{T\to\infty} I_T \leq I_{eq}. \tag{7.14}$$

What remains to be done is to prove, under appropriate assumptions that

$$\liminf_{T\to\infty} I_T \geq I_{eq}. \tag{7.15}$$

For a finite state Markov chain an appropriate assumption would be irreducibility.

If (7.14), (7.15) are correct and the Donsker-Varadhan assumptions also hold, then one must have $I_\infty = I_{eq}$ by (7.5). The fact that $I_{eq} \geq I_\infty$ follows easily from the following calculation. For any $g, I \in C(\Sigma)$, $g > 0$, we have by (6.7), (6.8)

$$\mu^g(k^g) = \mu^g(L^g I + k^g) \geq \mu^g[H(I)].$$

Now (7.3) can be rewritten as

$$\mathcal{J}(\mu) = \sup_I \mu[H(I)],$$

and hence

$$\mu^g(k^g) + \Psi(\mu^g) \geq \mathcal{J}(\mu^g) + \Psi(\mu^g),$$

$$I_{eq} \geq \inf_g [\mathcal{J}(\mu^g) + \Psi(\mu^g)] \geq I_\infty.$$

References

1. M. D. Donsker and S. R. S. Varadhan, Asymptotic evaluation of certain Markov process expectations for large time I, Comm, Pure Appl. Math. 27 (1975), 1-47.

2. Ibid Part II, Comm. Pure Appl. Math. 28 (1975), 279-301.
3. Ibid Part III, Comm. Pure Appl. Math. 29 (1976), 389-461.
4. L. C. Evans and H. Ishii, A PDE approach to some asymptotic problems concerning random differential equations with small noise intensities, preprint.
5. W. H. Fleming, Stochastic control for small noise intensities, SIAM J. Control 9 (1971), 473-517.
6. W. H. Fleming, Exit probabilities and stochastic control, Appl. Math. Optim. 4 (1978), 329-346.
7. W. H. Fleming and R. W. Rishel, Deterministic and Stochastic Optimal Control, Springer-Verlag, 1975.
8. W. H. Fleming and P. E. Souganidis, A PDE approach to asymptotic estimates for optimal exit probabilities, submitted to Annali Scuola Normale Superiore Pisa.
9. W. H. Fleming and M. Viot, Some measure-valued processes in population genetics theory, Indiana Univ. Math. J. 28 (1979), 817-843.
10. C. J. Holland, A new energy characterization of the smallest eigenvalue of the Schrödinger equation, Comm. Pure Appl. Math. 30 (1977), 755-765.
11. C. J. Holland, A minimum principle for the principal eigenvalue for second order linear elliptic equations with natural boundary conditions, Comm. Pure Appl. Math. 31 (1978), 509-520.
12. S.-J. Sheu, Optimal control and its application to large deviation theory, Brown Univ. PhD Thesis 1983.
13. S.-J. Sheu, Stochastic control and principal eigenvalue, Stochastics 11, (1984), 191-211.
14. M. I. Freidlin and A. D. Wentzell, Random Perturbations of Dynamical Systems, Springer-Verlag, 1984.

TOWARDS AN EXPERT SYSTEM IN STOCHASTIC CONTROL : OPTIMIZATION IN THE CLASS OF LOCAL FEEDBACKS

C.GOMEZ - J.P. QUADRAT - A.SULEM

I INTRODUCTION

Stochastic control problems can be solved completely or approximatively by different kind of approaches :

- dynamic programming
- decoupling technique
- stochastic gradient
- perturbation method.

The set of these methods are described in THEOSYS [11] for example.

For each approach we are designing a generator of program able to write automatically fortran program solving the problem.

In Gomez-Quadrat-Sulem [10] we have described a set of automatic tools to solve the problem by the dynamic programming approach.

In this paper we explain the decoupling approach, discuss the possibility of the corresponding generator. Then we give an example of generated program and the numerical results obtained by this generated program.

The plan is the following :

I. INTRODUCTION

II. OPTIMIZATION IN THE CLASS OF LOCAL FEEDBACKS

III. THE GENERATOR OF PROGRAM

IV. EXAMPLE

We want solve the stochastic control problem for diffusion processes that is

$$\min_{u} E \int_0^T C(t,X_t,U_t)dt$$

where U_t is the control and X_t is a diffusion process satisfying the stochastic differential equation

$$dX_t = b(t,X_t,U_t)dt + \sigma(t,X_t)dW_t$$

where W_t denotes a brownian motion b and σ are given functions.

When X_t belongs to $\mathbb{R}^n$ n large perhaps larger than 3 or 4 the traditional dynamic programming approach cannot be used practically. We have to apply other methods which do not give the optimal feedback but a good one or the optimum in a subsclass of the general feedback class.

In the next paragraph we explain the way of computing the optimal local feedback that is we suppose that each control is associated to a subsystem described by a subset I_i of the component of X_t and depends only of the corresponding components of the state.

$$U_i : (X_j,\ j \in I_i) \to R$$

$\bigcup_i I_i = \{1,\dots,n\}$ where n is the dimension of X.

II. OPTIMIZATION IN THE CLASS OF LOCAL FEEDBACKS.

In this paragraph we give the optimality conditions in the class of local feedbacks, and show that it is more difficult to solve these conditions than to compute the solution of the Hamilton-Jacobi equation. Then we study two particular cases :

- the case of the uncoupled dynamics,

- the case of systems having the product form property.

In these cases only it is possible to compute the optimal local feedbacks for large systems. Finally we discuss briefly the decoupling point of view.

2.1. The general situation.

Given I the indexes of the subsystems $I = \{1,2,\ldots,k\}$ n_i, [resp-m_i] denotes the dimension of the states [resp.the controls] of the subsystem $i \in I$. The local feedback S_i is a mapping of $\mathbb{R}^+ \times \mathbb{R}^{n_i}$ in $\mathcal{U}_i \subset \mathbb{R}^{m_i}$ the set of the admissible values of the control i. $\mathcal{S}_L$ denotes the class of local feedbacks $\mathcal{S}_L = \{S = (S_1,\ldots,S_k)\}$. Given the drift term of the system :

$$b : \mathbb{R}^+ \times \mathbb{R}^n \times \mathcal{U} \to \mathbb{R}^n$$
$$t \quad x \quad u \quad b(t,x,u)$$

with $\quad n = \sum_{i\in I} n_i, \mathcal{U} = \prod_{i\in I} \mathcal{U}_i,$

- the diffusion term :

$$\sigma : \mathbb{R}^+ \times \mathbb{R}^n \to M_n$$
$$t \quad x \quad \sigma(t,x)$$

with M_n the set of matrices (n,n) and $a = \frac{1}{2}\sigma\sigma^*$ where $*$ denotes the transposition

- the instantaneous cost :

$$c = \mathbb{R}^+ \times \mathbb{R}^n \times \mathcal{U} \to \mathbb{R}^+$$
$$t \quad x \quad u \quad c(t,x,u)$$

then boS [resp coS] denotes the functions $\mathbb{R}^+ \times \mathbb{R}^n \to \mathbb{R}^n$

$[\text{resp } \mathbb{R}^+ \times \mathbb{R}^n \to \mathbb{R}^+]$ $b(t,x,S(t,x))$ resp $c(t,x,S(t,x))$

Then if X^S denotes the diffusion (boS,a) (drift boS, and diffusion term σ) and P^S_μ its measure defined on $\Omega = C(\mathbb{R}^+, \mathbb{R}^n)$ with μ the law of the initial condition we want to solve

$$\underset{S \in \mathcal{S}_L}{\text{Min}} \; \mathbb{E}_{P^S_\mu} \int_0^T CoS(t,\omega_t)dt$$

where $\omega \in \Omega$, T denotes the time horizon. We have here a team of I players working to optimize a single criterion.

A simple way to obtain the optimality conditions is to consider another formulation of this problem : the control of the Fokker Planck equation that is :

$$\underset{S \in \mathcal{S}_L}{\text{Min}} \; J^S = \int_Q CoS(t,x)p^S(t,x)dt\, dx$$

with p solution of

$$\mathcal{L}_S^* p^S = 0$$

$$p^S(0,.) = \mu$$

with $Q = [0,T] \times \mathcal{O}$ and $\mathcal{O} = R^n$

$$\mathcal{L}_S = \frac{\partial}{\partial t} + \sum_j b_j oS \frac{\partial}{\partial x_j} + \sum_{i,j} a_{ij} \frac{\partial^2}{\partial x_i \partial x_j}$$

μ the law of the initial condition.

Then we have :

Theorem 1

A N.S.C. for $J^R \geq J^S$, R $S \in \mathcal{S}_L$, is that :

(1) $H(t,R,p^R,V^S) \geq H(t,S,p^R,V^S)$ pp in t

with

$$(2)\quad \begin{cases} \underline{H(t,R,p,V)} = \int_{\mathcal{O}} \underline{[CoR(t,x)} + \underline{\sum_i b_i oR(t,x) \frac{\partial V}{\partial x_i}} \underline{(t,x)]\, p(t,x)dx} \\ \underline{\mathcal{L}_R^* p^R = 0\ p^R(0,.) = \mu\ ; \mathcal{L}_S V^S + CoS = 0,\ V^S(T,.) = 0} \end{cases}$$

Remark 1. From this theorem the Pontriaguine condition can be obtained, that is a necessary condition of optimality of the strategy S is that : p,V,S satisfy

$$H(t,S,p^S,V^S) = \underset{R\in\mathcal{S}_L}{\mathrm{Min}}\ H(t,R,p^S,V^S)\ ;$$

$$(3)\quad \begin{cases} \mathcal{L}_S^* p^S = 0 \quad , \quad p(0,.) = \mu\ ; \\ \mathcal{L}_S V^S + CoS = 0 \quad , \quad V^S(T,.) = 0. \end{cases}$$

A proof is given in J.L. Lions [8].

Remark 2. This theorem give an algorithm to improve a given strategy R that is :

Step 1 : compute p^R

Step 2 : solve backward simultaneously

$$(4)\quad \begin{cases} \mathcal{L}_S V^S + CoS = 0 \qquad V^S(T,.) = 0 \\ S \in \underset{Z}{\mathrm{Arg\,Min}}\ H(t,Z,p^R,V^S) \end{cases}$$

By this way we improve the strategy.

A fixed point of the application $R \to S$ will satisfy the conditions (3).

We see that one iteration (4) of this algorithm is more expensive than the computation cost of the solution of the H.J.B. equation.

2.2. *Uncoupled dynamic systems.*

This is the particular case where b_i is a function of x_i and u_i, $\forall i \in I$

$$\begin{array}{rccccl} b_i : & \mathbb{R}^+ & \times\ \mathbb{R}^{n_i} & \times\ \mathcal{V}_i & \to & \mathbb{R}^{n_i} \\ & t & x_i & u_i & & b_i(t,x_i,u_i) \end{array}$$

and the noises are not coupled between the subsystems that is :

$$\sigma_i : \mathbb{R}^+ \times \mathbb{R}^{n_i} \to M_{n_i}$$
$$t \quad x_i \qquad \sigma_i(t,x_i)$$

In this situation we have

$$p^R = \prod_{i\in I} p_i^{R_i}$$

with $p_i^{R_i}$ solution of

(5) $$\mathcal{L}^*_{i,R_i} p_i^{R_i} = 0 \qquad p_i^{R_i}(0,.) = \mu_i \qquad \text{with} \quad = \prod_{i\in I} \mu_i$$

and

$$\mathcal{L}'_{i,R_i} = \frac{\partial}{\partial t} + \sum_{k\in I_i} b_k o R_i(t,X) \frac{\partial}{\partial X_k} + \sum_{k,\ell\in I_i} a_{k\ell} \frac{\partial^2}{\partial X_k \partial X_\ell}$$

with $$I_i = \{ \sum_{j<i} n_j < k \le \sum_{j<i+1} n_j \}.$$

Let us denote by

(6) $$C_i^R o R_i : \mathbb{R}^+ \times \mathbb{R}^{n_i} \to \mathbb{R}^+$$
$$t \quad x_i \qquad \int CoR(t,x) \prod_{j\neq i} p_j^{R_j}(t,x_j)dx_j$$

That is the conditional expectation of the instantaneous cost knowing the information only on the local subsystem i.

We have the following sufficient conditions to be optimal player by player :

Theorem 2. A sufficient condition for a strategy S to be optimal player by player is that the following conditions are satisfied :

(7) $$\underset{R_i}{\text{Min}} [\mathcal{L}_{i,R_i} V_i + C_i^R o R_i] = 0, \quad i \in I ;$$

with $C_i^R o R_i$ defined by (6) and (5)

The optimal cost is $\mu_1(V_1) .. = \mu_I(V_I)$ with $\mu_i(V_i) = \int_{\mathbb{R}^{n_i}} \mu_i(dx_i)V_i(o,x_i)$

Remark 3. The theorem 3 gives an algorithm to compute a feedback optimal player by player

given $\varepsilon, \nu \in \mathbb{R}^+$

Step 1) Choose $i \in I$

Solve (7)

if : $\mu_i(V_i) \leq \nu - \varepsilon$ than $\nu : \mu_i(V_i)$

$$R_i : = \text{Arg} \underset{R_i}{\text{Min}} \{\mathcal{L}_{i,R_i} V_i + C_i^R o R_i\}$$

if not choose another $i \in I$ until

$$\mu_i(V_i) \geq \nu - \varepsilon, \forall i \in I.$$

Step 2) When $\mu_i(V_i) \geq \nu - \varepsilon$, $\forall i \in I$, than $\varepsilon : = \frac{\varepsilon}{2}$, go to step 1.

By this algorithm we obtain a decreasing sequence $\nu^{(n)}$ which converges to a cost optimal player by player.

For a proof of a discrete version of this algorithm see Quadrat-Viot [1].

Remark 4. The interpretation of $V_i(t,X_i)$ $i \in I$ in terms of the variables of theorem 1 is :

$$V_i(t,x_i) = \int V(t,x) \prod_{j \neq i} p^{R_j}(t,X_j)dX_j$$

Remark 5. In this situation we have to solve a coupled system of P.D.E. but each of them is defined on a space of small dimension. By this way we can optimize, in the class of local feedback, systems which are not reachable by H.J.B. equation. An application to hydropower systems is given in Delebecque-Quadrat [2].

2.3. *Systems having the product form property.*

The property that a system has its dynamic uncoupled is very restrictive:in this paragraph, we show systems which have their invariant measure uncoupled, they are limit of network of queues of Jackson type. This property can be used to apply to them the results of 2.2. for the corresponding ergodic control problem that is :

$$\underset{S}{\text{Min}} \lim_{T\to\infty} \frac{1}{T} \int_0^T \text{CoS}(\omega_t)dt$$

Given B a generator of a Markov chain defined on $E = \{1,2,\ldots,n\}$, a function $\begin{array}{l} E \times R \to R \\ (i,x) \quad u_i(x) \end{array}$ a matrix $\sigma \in M_n$, $A = \frac{1}{2}\sigma\sigma^*$, D a diagonal matrix satisfying :

(8) $$DB^* + BD + 2A = 0$$

Theorem 3.

The invariant measure of probability p of the diffusion ($b = Bu$, $a=A$) such that (8) is true has the product form property that is :

(9) $$p(x) = C \prod_{i=1}^{n} p_i(x_i) \quad , \quad i \in E$$

(10) $$p_i(x_i) = \exp - \frac{1}{d_{ii}} \int_0^{x_i} u_i(s)ds$$

where C is a constant of normalization.

Demonstration : The Fokker-Planck equation can be written :

(11) $$- \text{div}\,[bp] + \text{div}\,[A \text{ grad } p] = 0$$

Let us make the change of variables $p = \exp V$ in (11), we obtain

$$(\text{grad } V, b - A \text{ grad } V) + \text{div}\,(b - A \text{ grad } V) = 0$$

Using (10), we have :

(12) $$(D^{-1}u, (B + AD^{-1})u) + \text{tr}\,[(B + AD^{-1}) \text{ grad } u] = 0$$

The quadratic part in (u) of (12) is equal to 0 if and only if :

$$D^{-1}B + B^*D^{-1} + 2D^{-1}AD^{-1} = 0$$

which can be written :

$$BD + DB^* + 2A = 0$$

which is (8).

We have also tr $[B + AD^{-1}]$ grad u = 0. Indeed grad u is diagonal because u_i is a function of x_i only and the coefficient of $\frac{\partial u_i}{\partial x_i}$ is $b_{ii} + a_{ii} / d_{ii}$ which is equal to zero by (8).

Remark 6. This class of diffusion processes are quite natural if we see them as the limit process when $N \to \infty$, obtained from Jackson network of queues by the scaling $x \to \frac{x}{N}$, $t \to \frac{t}{N^2}$.

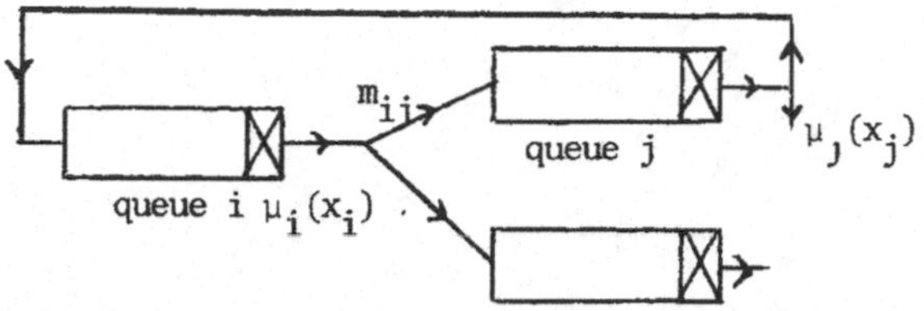

where $\mu_i(x_i)$ is the output rate of the queue i, m_{ij} is the probability of a customer leaving the queue i to go to the queue j.

The correlation of the noise given by (8) corresponds to system for which the noise satisfies a conservation law (for example the total number of customer in a closed network of queues).

Remark 7. We can now apply the result of 2.2 to compute the optimal local feedback for systems having the product form property and an ergodic criterion. Indeed :

$$\text{Min } \frac{1}{T}\int_0^T \text{CoS } (\omega_t)dt = \int \text{CoS } (x)\, p(x)dx$$

$$p(x) = \prod_{i=1}^{n} p_i(x_i)$$

and p_i satisfies :

$$-\frac{\partial}{\partial x_i}[u_i p_i] + \frac{\partial^2}{\partial x_i^2}[d_{ii}\, p_i] = 0, \qquad i \in E$$

$$\int p_i(x_i)dx_i = 1$$

2.4. *Remarks on decoupling.*

Another way to use the results of 2.2 when the dynamic is coupled is to do a change of feedback;let us consider the simpler case

$$b : \mathbb{R}^n \times U \to \mathbb{R}^n \qquad \text{with } u \in \mathbb{R}^n$$
$$x \quad u \quad b(x,u)$$

we use the feedback transformation $v = b(x,u)$ to decouple the drift terms. Now v is the control and we can apply the results of 2.2 to compute the best local feedback $v_i = S_i(x_i)$.

Then the solution in u of

(13) $\quad b(x,u) = S(x)$

gives the best feedback among the class that we can call "decoupling feedbacks".

One difficulty with this approach is for example the constraints on the control: the image by b of an hypercube is not in general an hypercube and if we take for constraints on the new control $v \in V(x) \subset b(x,\mathcal{U})$ with V(x) an hypercube of $\mathbb{R}^n$, the loss of optimality can become unacceptable.

This approach is well studied for deterministic linear and non linear systems Wonham [3], Isidori [4] and in the dynamic programming litterature Larson [5], Claude [6], Levine [7].

III. A GENERATOR OF PROGRAM FOR COMPUTING THE OPTIMAL LOCAL FEEDBACK.

It is difficult to write an efficient and general program to solve problem described in 2.2. Indeed each subsystem can have a special structure, special boundary conditions. Moreover each subsystem can have different space dimensions.

We have written in MACSYMA a program able to generate automatically a large class of such problems. Where MACSYMA is a language developped at MIT for formal calculus purpose.

The class is precisely described by the following grammar where we use a kind of Backus-Naur notation ("|" for the or, and "<Word>for a non terminal word)..

<local-feedback>::=<criterium>,
Σ(<initial-condition>,<subsystems>)
type

<criterium>::=ψ,d

ψ is the coupling function function from R to R
d(t) is a demand function of time

<subsystems>::=m,<subsystem-type>

$m \in \mathbb{N}$ denotes the number of subsystems having its structure described by <subsystem-type>

<subsystem-type>::=<domain>,<inside -condition>,
<boundary-condition>

<domain>::= $[0,1]^n \times [0,T]$
$n \in \mathbb{N}$

<boundary-condition>::= Σ <boundary-condition>,<boundary-element>
<boundary-element>

<boundary-element>::=<X_i>=0 | <X_i>=1 | t=T

<X_i>::=$X_1|X_2|\ldots|X_n$
<y>::=$(X_1,X_2,\ldots X_n,t)$
<boundary-condition>::=$V=f\,|\,\frac{\partial V}{\partial n}=f$

<inside-condition>::=<dynamic>,<local-cost>

<dynamic>::= $\frac{\partial V}{\partial t} + \sum_{i=1}^{n} b_i(\langle y\rangle) \frac{\partial}{\partial X_i} V + a_i(\langle y\rangle) \frac{\partial^2 V}{\partial X_j^2}$ |

$$\frac{\partial V}{\partial t} + \sum_{i=1}^{n} b_i(\langle y\rangle,\langle u\rangle) \frac{\partial}{\partial X_i} V + \sum_{i=1}^{n} a_i(\langle y\rangle,\langle u\rangle) \frac{\partial^2 V}{\partial X_i^2},$$

<constraints>

<u>::=$(u_1,u_2,\ldots,u_p)$

<constraints>::= $[\alpha,\beta]^p$

$\alpha \in \mathbb{R}$
$\beta \in \mathbb{R}$
$p \in \mathbb{N}$ the dimension of the control

<local-cost>::= $\phi(\langle y\rangle) \mid \phi(\langle y\rangle,\langle u\rangle)$

<initial-condition>::= $p(\langle y\rangle)$

$$p : \mathbb{R}^n \to R \text{ such that } \int_{\langle \text{domain} \rangle} p(dx)dx = 1 .$$

Moreover we have to specify to the generator the method of discretization in time: explicit or implicit, in space, the method of optimization : newton, gradient, gradient with projection and so on.

With these informations the generator is able to write a Fortran program solving the problem. An example is given in the following chapter.

In the future we shall extend the class of systems that the generator is able to solve by generalizing :

- the structure of ψ,

- extending the method to ergodic and static problem,

- generalizing the structure of the control space,

- improving the numerical method of integration.

For the classical HJB equation a more general generator exists and is described in Gomez-Quadrat-Sulem [10].

IV. AN EXAMPLE

Let us consider the following stochastic control problem which models a water storage management for electricity generation.

The dynamics of the water stocks are the following :

$$dX_t^i = (a_t^i - u_t^i)dt + \sigma_i\, dw_t^i - d\xi_1^i + d\xi_2^i$$

$$0 \le u_t^i \le \sqrt{X_t^i}$$

where

- t denotes the time,

- i denotes index of a dam $i \in \{1,2,3\}$,

- X_t^i denotes the amount of water in the stock,

- a_t^i denotes the average imput of water,

- $\sigma_i dw_t^i$ the stochastic perturbation on the imput of water,

- ξ_1^i is an increasing process strictly increasing only when $X_t^i = 1$ denoting the cumulated overflowing water,

- ξ_2^i is an increasing process strictly increasing only when $X_t^i = 0$, denoting the cumulate water that we have to add to the imput in such way that X_t^i be always positive. It can be seen as a model correction indeed $a^i dt + \sigma_i dw_t^i$ is not almost surely positive but $a_t^i dt + \sigma_i dw_t^i + d\xi_2^i$ will be always positive when $X_t^i = 0$. By this way X_t^i is always positive.

The influence of ξ_2^i is small because $a_t^i > 0$ and $u_t^i = 0$ if $X_t^i = 0$,

The criterion is

$$E \int_0^T \psi(z_t - \sum_{i=1}^{3} u_t^i)$$

where

- z_t denotes the demand in electricity

- the function $\psi : x \to x^2$ denotes the generation cost of thermal means. Indeed $z_t - \sum_{i=1}^{z} u_t^i$ can be seen as the thermal electricity generation to be produced.

The following annex and figures show :

- the macsyma program which specifies the problem and calls the generator of fortran program

- the program generated,

- the main program calling the subroutine generated,

- the optimal price of water obtained by the local feedback method,

- the optimal price obtained by solving the complete HJB equation.

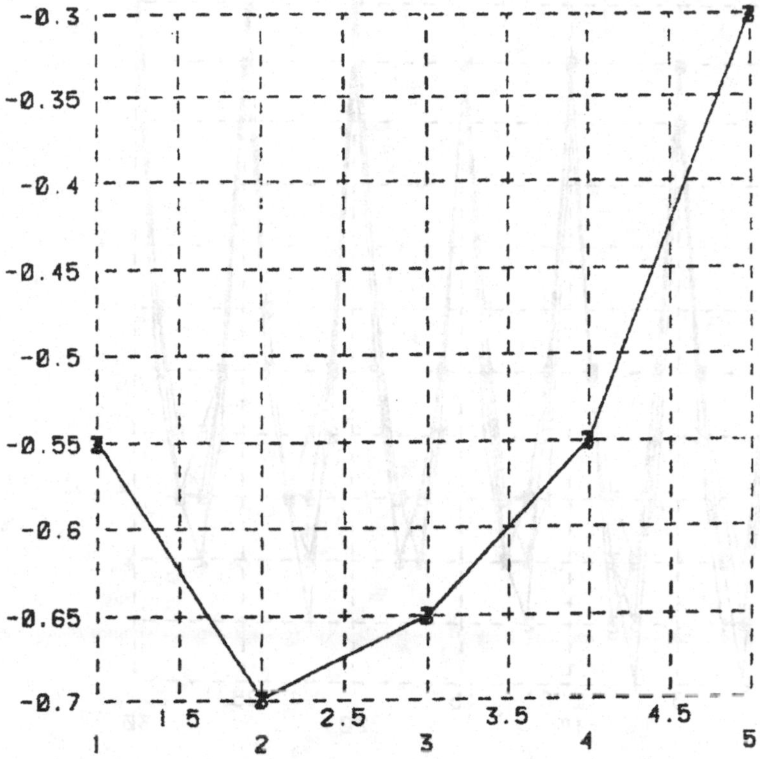

Figure 1

Minus the price of water $\frac{\partial v_1}{\partial x_1}$ as a function of the water level for one dam.

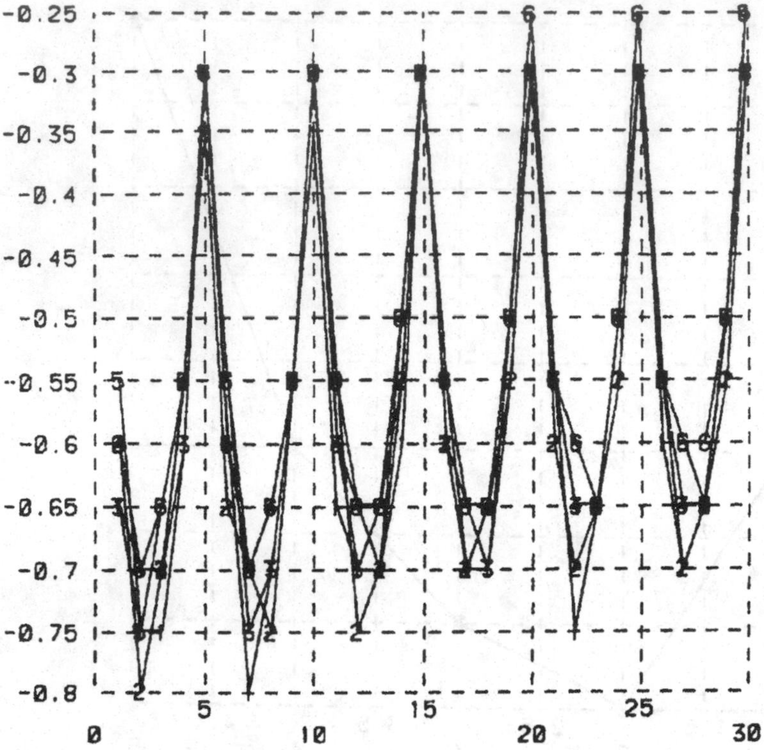

Figure 2

Minus the price of water ($\frac{\partial v}{\partial x}$) as a function of the three dimensional space obtained by solving the complete HJB equation. $X_2 = 0.1$ is represented by the abscisse 1 to 5 the section $X_2 = 0,3$ by the abscisse 5 to 10 etc...

ANNEX 1

Subroutine in macsyma specifying the control problem by the list "syst" and calling the generator of fortran program here "feedloc". In the future we shall use a semi-natural language interface to specify the problem.

```
appel():=(
     cline("dl belman.fortran"),
     cdl0:((nat,0)),
     cdl1:((nat,0)),
     ap:(1.0+cos(zf+44.0*x0/7.0))/2.0,
     demm:5.0+3.0*cos(44.0*x0/7.0)/2.0,
     hm:(ap-u1)*p1-u1,
     type:(1,parab,exp,0.0,pasecriture,pasmoy,condlim,cdl0,cdl1,parae,difu,
         derive,(0),plus,dif,(za),belm,1,newto,gradproj,hm,((0.0,zu*x^(1/2))),
          param,(zu,za,zf)),
     psi(x):=x^2,
     syst:(psi,demm,(1.0),(3,type)),
     feedloc(syst) )$
```

ANNEX 2

Subroutines fortran, automatically generated, solving the problem

```
      subroutine primal1(n1,n0,h0,v,u,eps,nmax,ymoen,variance,zu,za,zf,r
      1   og)
       dimension v(n1,n0),u(1,n1,n0),ymoen(n0),variance(n0)
c
c      Resolution de l equation de Bellman dans le cas ou:
c             Les parametres sont zu za zf
c             L etats-temps est:  x1 x0
```

```
c         La dynamique du systeme est decrite par l operateur
c                            p1 cos(zf + 6.2857142 x0)
c     plus( q1 za , Minu( ------------------------- - p1 u1
c                                      2
c
c                                                  2                        2
c (- u1 + ymoen(i0) + 2 variance(i0))    (- u1 + ymoen(i0) + variance(i0))
c+ ----------------------------------- + ---------------------------------
c                16                                    4
c
c                                              2                            2
c (- u1 + ymoen(i0) - variance(i0))    (- u1 + ymoen(i0) - 2 variance(i0))
c+ --------------------------------- + -----------------------------------
c               4                                      16
c
c                 2
c 3 (ymoen(i0) - u1)    p1
c+ ------------------ + -- ) )
c       8                2
c         ou v(..) et w designe  le cout optimal
c         ou pi designe sa derivee premiere par rapport a xi
c         ou qi designe sa derivee seconde par rapport a xi
c         Le probleme est parabolique
c         Le temps note x0 appartient a (0,(n0-1)*h0)
c         le cout sur l'etat final 0.0
c         Les conditions aux limites sont:
c              x1 = 0    -p1 = 0
c              x1 = 1    p1 = 0
c     Les nombres de points de discretisation sont: n1 n0
c              x1 = 1 correspond a i1 = n1 - 1
c              x1 = 0 correspond a i1 = 2

c     Le schema de discretisation en temps est explicite
c     p1 est discretise par difference divise  symetrique
c     Minimisation par la methode de  gradient avec projection
c
c                                                           de l'Hamiltonien:
c         p1 cos(zf + 6.2857142 x0)
c         ------------------------- - p1 u1
c                     2
```

```
c
c                                          2                                      2
c  (- u1 + ymoen(i0) + 2 variance(i0))     (- u1 + ymoen(i0) + variance(i0))
c+ -------------------------------------- + ----------------------------------
c                   16                                        4
c
c                                        2                                        2
c  (- u1 + ymoen(i0) - variance(i0))     (- u1 + ymoen(i0) - 2 variance(i0))
c+ ---------------------------------- + -------------------------------------
c                  4                                         16
c
c                      2
c  3 (ymoen(i0) - u1)     p1
c+ ------------------- + --
c          8               2
c      contraintes sur le controle:
c            0.0 =< u1 =< sqrt(x1) zu
c      nmax designe le nombre maxi d iteration de la methode de
c                                                      gradient avec projection
c      eps designe l erreur de convergence de la methode de
c                                                      gradient avec projection
c
      h1 = 0.999999/(n1-3)
      u1 = u(1,1,1)
      hih1 = h1**2
      h21 = 2*h1
      nm0 = n0-1
      nm1 = n1-1
      do  111   i1 = 1 , n1 , 1
      x1 = h1*(i1-2)
      v(i1,n0) = 0.0
  111 continue
      do  100   ii0 = 1 , nm0 , 1
      i0 = n0-ii0
      x0 = h0*(i0-1)
      v(n1,i0+1) = v(n1-1,i0+1)
      v(1,i0+1) = v(2,i0+1)
  110 continue
      do  109   i1 = 2 , nm1 , 1
```

```
      x1 = h1*(i1-2)
      q1 = (v(i1+1,i0+1)-2*v(i1,i0+1)+v(i1-1,i0+1))/hih1
      p1 = (v(i1+1,i0+1)-v(i1-1,i0+1))/h21
      niter = 0
      w0 = -1.0e+20
101   continue
      niter = niter+1
      if ( niter - nmax )  102 , 102 , 103
103   continue
      write(8,901)i1,i0
901   format(' descente n a pas converge', 2 i3)
      goto  104
102   continue
      un1 = (1-2*rog)*u1+(p1+2*ymoen(i0))*rog
      u1 = un1
      u1 = amax1(u1,0.0)
      u1 = amin1(u1,sqrt(x1)*zu)
      ww = p1*cos(zf+6.2857142*x0)/2.0-p1*u1+(-u1+ymoen(i0)+2*variance(i
     1   0))**2/16.0+(-u1+ymoen(i0)+variance(i0))**2/4.0+(-u1+ymoen(i0)-
     2   variance(i0))**2/4.0+(-u1+ymoen(i0)-2*variance(i0))**2/16.0+3.0
     3   *(ymoen(i0)-u1)**2/8.0+p1/2.0
      er = abs(ww-w0)
      if ( er - eps )  104 , 104 , 105
105   continue
      w0 = ww
      goto  101
104   continue
      u(1,i1,i0) = u1
      w0 = ww
      w1 = q1*za
      w0 = w1+w0
      vnew = h0*w0+v(i1,i0+1)
      v(i1,i0) = vnew
109   continue
100   continue
      return
      end
```

```
      subroutine dual1(n1,n0,h0,v,variance,ymoen,u,zu,za,zf)
      dimension variance(n0),ymoen(n0),v(n1,n0),u(1,n1,n0)
c
c      Resolution de l equation de Fokker_Planck dans le cas ou:
c            Les parametres sont zu za zf
c            L etats-temps est:  x1 x0
c            La dynamique du systeme est decrite par l operateur
c          2
c         d           d        cos(zf + 6.2857142 x0)         1
c        ---- (v za) - --- (v (----------------------- - u1 + -))
c          2          dx1                2                    2
c       dx1
c            ou v(..) et w designe  la densite de probabilite
c            Le probleme est parabolique
c            Le temps note x0 appartient a (0,(n0-1)*h0)
c            la condition initiale 1.0
c            variance designe la variance de - u1
c            ymoen designe la moyenne de - u1
c            Les conditions aux limites sont:
c                             d               cos(zf + 6.2857142 x0)         1
c                 x1 = 0     --- (v za) - v (----------------------- - u1 + -) = 0
c                            dx1                        2                    2
c
c
c                               cos(zf + 6.2857142 x0)         1     d
c                 x1 = 1     v (----------------------- - u1 + -) - --- (v za) = 0
c                                          2                    2    dx1
c
c
c      Les nombres de points de discretisation sont: n1 n0
c                 x1 = 1 correspond a i1 = n1 - 1
c                 x1 = 0 correspond a i1 = 2
c      Le schema de discretisation en temps est explicite
c
      h1 = 0.999999/(n1-3)
      hih1 = h1**2
      nm0 = n0-1
      nm1 = n1-1
      x0 = 0
```

```
      do  106   i1 = 1 , n1 , 1
      x1 = h1*(i1-2)
      v(i1,1) = 1.0
  106 continue
      do  100   i0 = 2 , n0 , 1
      x0 = h0*(i0-1)
      v(n1,i0-1) = v(n1-1,i0-1)*(h1*(cos(zf+6.2857142*x0)/2.0-u(1,n1-1,i
     1 0-1)+1.0/2.0)/2.0+za)
      v(1,i0-1) = v(2,i0-1)*(za-h1*(cos(zf+6.2857142*x0)/2.0-u(1,2,i0-1)
     1  +1.0/2.0)/2.0)
  103 continue
      do  102   i1 = 2 , nm1 , 1
      x1 = h1*(i1-2)
      p1 = v(i1+1,i0-1)*(za/hih1-(cos(zf+6.2857142*x0)/2.0-u(1,i1+1,i0-1
     1  )+1.0/2.0)/h1/2.0)+v(i1-1,i0-1)*((cos(zf+6.2857142*x0)/2.0-u(1,
     2  i1-1,i0-1)+1.0/2.0)/h1/2.0+za/hih1)-2*v(i1,i0-1)*za/hih1
        if (i1.eq.nm1) p1 = p1-v(n1,i0-1)*(-0.5*(0.5*cos(zf+6.2857142*x0
     1  )-u(1,n1,i0-1)+0.5)/h1+za/hih1-1/hih1)
        if (i1.eq.2) p1 = v(1,i0-1)*(-0.5*(0.5*cos(zf+6.2857142*x0)-u(1,
     1  1,i0-1)+0.5)/h1-za/hih1+1/hih1)+p1
      w0 = p1
      vnew = h0*w0+v(i1,i0-1)
      v(i1,i0) = vnew
  102 continue
      ymo1 = 0.0
      ymo2 = 0.0
      do  104   i1 = 2 , n1 - 1 , 1
      x1 = h1*(i1-2)
      ymo1 = ymo1-u(1,i1,i0-1)*v(i1,i0-1)/(n1-2)
      ymo2 = ymo2+u(1,i1,i0-1)**2*v(i1,i0-1)/(n1-2)
  104 continue
      ymoen(i0-1) = ymo1
      variance(i0-1) = ymo2-ymo1**2
  100 continue
      return
      end
```

```
      subroutine fedloc(n11,n0,ymoen,variance,dem,vdem,h0,dem1,vdem1,u1,
     1   vv1,v1,pr1,nflmax,epsilon,epsimp,impmax,eps,nmax,ro1,rog1)
      common/parametre/zu(3),za(3),zf(3)
      dimension ymoen(n0),variance(n0),dem(n0),vdem(n0),dem1(3,n0),vdem1
     1   (3,n0),u1(1,n11,n0),vv1(3,n11,n0),v1(n11,n0),pr1(n11,n0)
c
c          Optimisation dans la classe des feedbacks locaux d'un
c        systeme compose de sous-systemes a dynamiques decouplees
c        mais couples par le critere
c          Il y a 1 types de sous-systemes
c            - 3 sous systeme de type 1
c          Les sous systemes sont decrits precisement dans les
c        commentaires des sous-programmes primaux et duaux corres
c        pondants
c                                          2
c          Le critere s'ecrit: (p + d)
c          d designe la demande : 1.5 cos(6.2857142 x0) + 5.0
c          p la production somme des productions locales pi
c          pi designe la production d'un sous-ystme de type i
c            p1 = - u1
c          Le critere est evalue pour la condition initiale:
c            v1 = 1.0
c        ou vi designe la densite de probabilite initiale des
c        sous-systeme de type i
c          La methode de resolution est une methode de relaxation
c          Les parametres d'appel sont:
c            -epsilon l'erreur de convergence de la relaxation entre
c        sous systeme
c            -nflmax le nbre maxi d'iterations correspondantes
c            -epsimp l'erreur de cvgce pour les systemes implicites
c            -impmax le nbre d'iteration maxi correspondante
c            -rogi controle la conver. de la meth. de desc. du syst i
c            -roi controle la convergence du syst implicite i
c            -eps l'erreur de convergence dans la methode de Newton
c            -nmax le nbre maxi d'iterations correspondantes
c          Le temps note x0 appartient a (0,(n0-1)*h0)
c          nij designe le nbre de pts de discretisation de la
c        composante i d'un sous systeme de type j
```

```
c          Les sorties sont:
c            -vvi(j,...) designe le cout vu par le j-eme sous-systeme
c         de type i
c            -demi(j,..) la production moyenne correspondante
c            -vdemi(j,..) la variance de la production correspondante
c          Les autres parametres ne servent que pour avoir des
c         dimensions variables dans le sous programmes
c          Les parametres  (zu, za, zf) doivent etre passes dans le
c         common parametre
c
      do  100   i0 = 1 , n0 , 1
      x0 = h0*(i0-1)
      dem(i0) = 1.5*cos(6.2857142*x0)+5.0
  100 continue
      do  101   j = 1 , 3 , 1
      call dual1(n11,n0,h0,pr1,variance,ymoen,u1,zu(j),za(j),zf(j))
      do  102   i0 = 1 , n0 , 1
      dem1(j,i0) = ymoen(i0)
      dem(i0) = ymoen(i0)+dem(i0)
      vdem1(j,i0) = variance(i0)
      vdem(i0) = vdem(i0)+variance(i0)
  102 continue
  101 continue
      coutv = 10000000000
      nitfl = 0
      write(8,901)
  901 format('  converg:(')
  113 continue
      nitfl = nitfl+1
      do  103   j = 1 , 3 , 1
      do  104   i0 = 1 , n0 , 1
      dem(i0) = dem(i0)-dem1(j,i0)
      vdem(i0) = vdem(i0)-vdem1(j,i0)
  104 continue
      call primal1(n11,n0,h0,v1,u1,eps,nmax,dem,vdem,zu(j),za(j),zf(j),r
     1   og1)
      do  105   i1 = 1 , n11 , 1
      do  105   i0 = 1 , n0 , 1
      vv1(j,i1,i0) = v1(i1,i0)
  105 continue
```

```
      coutneuf = 0
      do  106   i1 = 2 , n11 - 1 , 1
      x1 = 0.999999*(i1-2)/(n11-3)
      coutneuf = v1(i1,1)/(n11-2)+coutneuf
  106 continue
      write(8,902)coutneuf
  902 format('  ',e14.7,',')
      call dual1(n11,n0,h0,pr1,variance,ymoen,u1,zu(j),za(j),zf(j))
      do  107   i0 = 1 , n0 , 1
      dem1(j,i0) = ymoen(i0)
      dem(i0) = ymoen(i0)+dem(i0)
      vdem1(j,i0) = variance(i0)
      vdem(i0) = vdem(i0)+variance(i0)
  107 continue
  103 continue
      if ( nflmax - nitfl )  110 , 109 , 109
  110 continue
      write(8,900)
  900 format(' feedloc n a pas converge')
      goto  112
  109 continue
      if ( - epsilon + coutv - coutneuf )  112 , 112 , 111
  111 continue
      coutv = coutneuf
      goto  113
  112 continue
      write(8,903)
  903 format(' ())$')
      return
      end
```

ANNEX 3

Main program to write by hand to call the subroutine feedloc which solves the problem

```
   dimension dem1(3,61),vdem1(3,61),vv1(3,13,61)
      dimension ymoen(61),variance(61),dem(61),vdem(61),v1(13,61),
     1 u1(1,13,61),pr1(13,61)
      common /parametre/zu(3),za(3),zf(3)
      do 100 i=1,61
      do 100 j=1,3
      u1(1,j,i)=1.0
100   continue
      do 101 j=1,3
      za(j)=0.18
      zf(j)=1.57
      zu(j)=3.0
101   continue
      call fedloc(13,61,ymoen,variance,dem,vdem,0.009,dem1,vdem1,u1,
     1 vv1,v1,pr1,10,.01,0.01,100,0.01,20,0.01,0.5)
      write (9,200)
200   format (" v:(")
      do 202 jj=1,6
      j=1+10*(jj-1)
      write (9,201)((vv1(k,i,j),i=1,13),k=1,3)
201   format(" (",38(f4.2,","),f4.2,"),")
202   continue
      write (9,203)
203   format(" ())$")
      stop
      end
```

REFERENCES.

[1] QUADRAT - VIOT : Product form and optimal local feedback for a multiindex Markov Chain, 18th Allerton Conference, October 1980.

[2] DELEBECQUE - QUADRAT : Contribution of stochastic control singular perturbation team theories to an example of large scale system : management of hydropower production, IEEE AC, April 1978, pp. 209-222.

[3] WONHAM : Linear system : geometric approach, Springer Verlag, 1974.

[4] ISIDORI : The geometric approach to non linear feedback control : a survey, 5th Conference on "Analyse et Optimisation des Systèmes", Versailles, 1982, Lecture Notes in Control and Information Sciences n°44, Springer Verlag.

[5] LARSON - KORSAK : A dynamic programming successive approximations : technique with convergence proofs Part I & II, Automatica, 1969.

[6] CLAUDE : Linéarisation par difféomorphisme et immersion des systèmes, 6th Conference "Analyse et Optimisation des Systèmes", Nice, June 1984, Springer Verlag, Lecture Notes in Control and Information Sciences

[7] GEROMEL - LEVINE - WILLIS : A fast algorithm for systems decoupling using formal calculus, 6th Conference "Analyse et Optimisation des Systèmes", Nice, Juin 1984, Springer Verlag, Lect. Notes in Control and Information Sciences.

[8] J.L. LIONS : Contrôle optimal des systèmes gouvernés par des équations aux dérivées partielles, Paris, Dunod 1968.

[9] Mit Mathlab Group : MACSYMA, Manual, Mit Press.

[10] GOMEZ - QUADRAT - SULEM : Vers un système expert en contrôle stochastique, 6th Conference "Analyse et Optimisation des Systèmes", Nice, Juin 1984, Springer Verlag, Lecture Notes in Control and Information Sciences.

[11] THEOSYS, Commande Optimale de systèmes stochastiques, RAIRO Automatique, à paraître.

OPTIMAL CONTROL AND VISCOSITY SOLUTIONS

P.L. Lions
Ceremade
University Paris-Dauphine
Place de Lattre de Tassigny
75775 Paris Cedex 16

Introduction:

M. G. Crandall and the author have introduced recently (see [8], [9], and M.G. Crandall , L.C. Evans and P.L. Lions [7]) the notion of viscosity solutions of first-order Hamilton-Jacobi equations and proved general uniqueness and comparison results. The corresponding existence results may be found in P.L. Lions [16], [17], P.E. Souganidis [24], G. Barles [1], [2]. We briefly recall the definition of viscosity solutions and the above results in sections I, II.

In section III below, we show the intrinsic relations between optimal deterministic control problems (and the dynamic programming principle) and viscosity solutions of the corresponding Hamilton-Jacobi equations (those equations in engineering literature are often called Bellman equations). These relations are taken from P.L. Lions [16], [18], [19] (see also P.L. Lions and M. Nisio [22]). We also recall in section III the "everywhere formulation" of Bellman equation in the case when the value function is locally Lipschitz: this formulation in its general form is due to P.L. Lions and P.E. Souganidis [23], a weaker form appearing first in L.C. Evans and H. Ishii [11].

Some applications of the use of viscosity solutions in control problems may be found in I. Capuzzo-Dolcetta [5], I. Capuzzo-Dolcetta and L.C. Evans [6], G. Barles [3], [4], L.C. Evans and H. Ishii [12], W.H. Fleming and P.E. Souganidis [14] ...

Section IV is devoted to a typical example of infinite horizon deterministic control problem with state constraints: we give a characterization of the value function using viscosity solutions. Let us mention that the results of sections III, IV are easily extended to differential games.

Finally in section V, we report briefly on the extension of the notion of viscosity solutions to fully nonlinear, degenerate elliptic, second-order equations and their relations with optimal stochastic control. These results are taken from P.L. Lions [18], [19], [20].

Summary

I. Viscosity solutions: definition and elementary properties.
II. Existence, uniqueness and comparison results.
III. Viscosity solutions and optimal deterministic control.
IV. State-constraints problems.
V. Extension to second-order equations.

I. Viscosity solutions: definition and elementary properties.

Let $\mathcal{O}$ be an open set in $\mathbb{R}^N$, we define below viscosity solutions of

(HJ) $H(x,u,Du) = 0$ in $\mathcal{O}$

where $H(x,t,p) \in C(\mathcal{O} \times \mathbb{R} \times \mathbb{R}^N)$, u is scalar unknown function - is scalar. We first recall that if $v \in C(\mathcal{O})$, the superdifferential (resp. subdifferential) of v at $x \in \mathcal{O}$, denoted by $D^+v(x)$ (resp. $D^-v(x)$), is the closed convex set (possibly empty), defined by:

(1) $$D^+v(x) = \{\xi \in \mathbb{R}^N \mid \limsup_{y \to x, y \in \mathcal{O}} \{v(y)-v(x)-(\xi, y-x)\}|y-x|^{-1} \leq 0\}$$

$$\text{(resp. } D^-v(x) = \{\xi \in \mathbb{R}^N \mid \liminf_{y \to x, y \in \mathcal{O}} \{v(y)-v(x)-(\xi, y-x)\}|y-x|^{-1} \geq 0\})$$

or equivalently

(2) $$D^+v(x) = \{\xi \in \mathbb{R}^N \mid \phi \in C^1(\mathcal{O}), \phi(x)=v(x), D\phi(x)=\xi \text{ and } \phi(y) > v(y) \text{ if } y \in \mathcal{O},\ y \neq x\}$$

$$\text{(resp. } D^-v(x) = \{\xi \in \mathbb{R}^N \mid \phi \in C^1(\mathcal{O}), \phi(x)=v(x), D\phi(x)=\xi \text{ and } \phi(y) < v(y) \text{ if } y \in \mathcal{O},\ y \neq x\}).$$

We may define now viscosity solutions of (HJ):

Definition: Let $u \in C(\mathcal{O})$; u is said to be a viscosity subsolution (resp. supersolution) of (HJ) if

(3) $H(x,u(x),\xi) \leq 0$, $\forall x \in \mathcal{O}$, $\forall \xi \in D^+u(x)$;

(resp. (4) $H(x,u(x),\xi) \geq 0$, $\forall x \in \mathcal{O}$, $\forall \xi \in D^-u(x)$).

Finally u is a viscosity solution of (HJ) if u is a viscosity subsolution and a viscosity supersolution of (HJ). □

Remarks: 1) Of course if $D^+u(x) = \emptyset$, we do not have to check (3) at x.
2) If $u \in C^1(\mathcal{O})$ solves (HJ) then, observing that $D^+u(x)=D^-u(x)=\{Du(x)\}$ for all $x \in \mathcal{O}$, u is clearly a viscosity solution of (HJ).
3) If u is a viscosity solution of (HJ) and if u is differentiable at $x_0 \in \mathcal{O}$ then $D^+u(x_0) = D^-u(x_0)=\{Du(x_0)\}$ and thus

$$H(x_0,u(x_0),Du(x_0)) = 0 \quad . \quad \square$$

For piecewise C^1 functions, the proposition below gives a very simple criterion for a function to be a viscosity solution: let us assume that $\mathcal{O}$ is divided in two components $\mathcal{O}_1$, $\mathcal{O}_2$ by a smooth hypersurface Γ ($\mathcal{O} = \mathcal{O}_1 \cup \mathcal{O}_2 \cup \Gamma$, $\mathcal{O}_1 \cap \mathcal{O}_2 = \emptyset$) and that $u = u_1$ on $\mathcal{O}_1$, $u = u_2$ on $\mathcal{O}_2$ where $u_1 \in C^1(\mathcal{O}_1 \cup \Gamma)$, $u_2 \in C^1(\mathcal{O}_2 \cup \Gamma)$, $u_1 = u_2$ on Γ (and we set $u = u_1 = u_2$ on Γ). For any $x \in \Gamma$, we denote by T_x the tangent hyperplane to Γ at x, by n(x) the unit normal to Γ at x pointing into $\mathcal{O}_2$, and we decompose $\xi \in \mathbb{R}^N$:

$$\xi = P_x \xi + (\xi,n(x))n(x) \ , \quad P_x\xi \in T_x \ .$$

Clearly we have: $P_x(Du_1(x)) = P_x(Du_2(x))$, $\forall x \in \Gamma$.
We observe next that the following holds at any $x \in \Gamma$:

i) if $\frac{\partial u_1}{\partial n}(x) \geq \frac{\partial u_2}{\partial n}(x)$, $D^+u(x)=\{\xi=P_x(Du_1(x))+\lambda n(x) / \lambda \in [\frac{\partial u_2}{\partial n}(x),\frac{\partial u_1}{\partial n}(x)]$

and $D^-u(x)=\emptyset$ if $\frac{\partial u_1}{\partial n}(x) > \frac{\partial u_2}{\partial n}(x)$, $D^-u(x)=\{Du_1(x)\}=\{Du_2(x)\}$ if

$$\frac{\partial u_1}{\partial n}(x) = \frac{\partial u_2}{\partial n}(x);$$

ii) if $\frac{\partial u_1}{\partial n}(x) < \frac{\partial u_2}{\partial n}(x)$, $D^-u(x)=\{\xi=P_x(Du_1(x))+\lambda n(x) / \lambda \in [\frac{\partial u_1}{\partial n}(x), \frac{\partial u_2}{\partial n}(x)]\}$

and $D^+u(x) = \emptyset$

Finally we assume that u_1, u_2 solve (HJ) respectively in $\mathcal{O}_1$, $\mathcal{O}_2$. Applying the definition and the above characterizations we find the:

Proposition 1: With the above notations and assumptions, u is a viscosity subsolution (resp. supersolution) if we have:

$$(5)\quad \begin{cases} H(x,u(x),\theta Du_1(x) + (1-\theta)Du_2(x)) \le 0 \ , \quad \forall \theta \in [0,1] \ , \\ \text{for all } x \in \Gamma \text{ such that: } \frac{\partial u_1}{\partial n}(x) < \frac{\partial u_2}{\partial n}(x) \ ; \end{cases}$$

(resp.)

$$(6)\quad \begin{cases} H(x,u(x),\theta Du_1(x) + (1-\theta)Du_2(x)) \ge 0 \ , \quad \forall \theta \in [0,1] \ , \\ \text{for all } x \in \Gamma \text{ such that: } \frac{\partial u_1}{\partial n}(x) < \frac{\partial u_2}{\partial n}(x) \ . \end{cases}$$

) □

Remarks: 1) Observe that for all $x \in \Gamma$: $H(x,u(x),Du_i(x))=0$, $\forall i = 1,2$.
2) if H is convex in p (resp. concave), (5) (resp. (6)) trivially holds and u is always a viscosity subsolution (resp. supersolution). □

Using the formulation (2) of D^+ and observing that if $u \in C(\mathcal{O})$, $\phi \in C^1(\mathcal{O})$ and $u-\phi$ has a local maximum at $x \in \mathcal{O}$ then

$$D\phi(x) \in D^+u(x)$$

(indeed for y near x, $u(y) \le \phi(y)+u(x)-\phi(x)=u(x)+(D\phi(x),y-x)+O(|y-x|)$) we immediately obtain the following equivalent definition of viscosity solutions:

Proposition 2: Let $u \in C(\mathcal{O})$; u is a viscosity subsolution (resp. supersolution, resp. solution) of (HJ) if and only if we have for all $\phi \in C^1(\mathcal{O})$:

(7) at each local maximum point x of $u - \phi$, $H(x,u(x),D\phi(x)) \le 0$

(8) (resp. at each local minimum point x of $u - \phi$, $H(x,u(x),D\phi(x)) \ge 0$; resp. (7) and (8) hold). □

Remark: Clearly enough we may replace C^1 by C^2 or C^∞, local by global, or local strict, or even global strict. □

This formulation has the advantage of yielding the

Proposition 3: i) Let $u_n \in C(\mathcal{O})$ be a viscosity subsolution (resp. supersolution, resp. solution) of

$$H_n(x,u_n,Du_n) = 0 \text{ in } \mathcal{O}.$$

Let us assume that u_n, H_n converge uniformly on compact subsets of $\mathcal{O}$, $\mathcal{O} \times \mathbb{R} \times \mathbb{R}^N$ to u,H. Then u is a viscosity subsolution (resp. supersolution, resp. solution) of (HJ).

ii) Let $u_n \in C^2(\mathcal{O})$ satisfy:

$$-\varepsilon_n \Delta u_n + H_n(x,u_n,Du_n) \leq 0 \text{ in } \theta \text{ (resp. } \geq 0, \text{ resp. } = 0)$$

and assume that $\varepsilon_n > 0$, $\varepsilon_n \underset{n}{\to} 0$, u_n, H_n converge uniformly on compact subsets of $\mathcal{O}$, $\mathcal{O} \times \mathbb{R} \times \mathbb{R}^N$ to u, H. Then u is a viscosity subsolution (resp. supersolution, resp. solution) of HJ. □

<u>Proof</u>: The proof of i) and ii) are very similar, thus we will only prove ii) and the "subsolution" part of it. In view of Proposition 2 and the remark following it, we have to check (7) at a local strict maximum point x_0 of $u - \phi$ where $\phi \in C^2(\mathcal{O})$. Since u_n converges uniformly on compact subsets of $\mathcal{O}$ to u, we deduce that for n large $u_n - \phi$ admits admits a local maximum point x_n which converges to x_0 as $n \to \infty$. Next, $Du_n(x_n) = D\phi(x_n)$, $\Delta u_n(x_n) \leq \Delta\phi(x_n)$ and thus

$$H_n(x_n,u_n(x_n), D\phi(x_n) \leq \varepsilon_n \Delta u_n(x_n) \leq \varepsilon_n \Delta\phi(x_n)$$

and we conclude letting n go to + ∞. □

The last result we wish to mention is the

<u>Proposition 4</u>: i) Let $u \in W^{1,\infty}_{loc}(\mathcal{O})$ satisfy:

$$H(x,u,Du) \leq 0 \quad \text{a.e. in } \mathcal{O}$$

and assume that H(x,t,p) is convex in p for all $(x,t) \in \mathcal{O} \times \mathbb{R}$; then u is a viscosity subsolution of (HJ).

ii) Let $u \in W^{1,\infty}_{loc}(\mathcal{O})$ satisfy:

$$H(x,u,Du) \geq 0 \text{ a.e. in } \mathcal{O}, \quad \partial^2_\xi u \geq -C \text{ in } \mathcal{D}'(\mathcal{O}), \quad \forall |\xi| = 1$$

for some constant C, then u is a viscosity supersolution of (HJ). □

Remark: Part i) still holds if we replace the convexity of H in p by

$$\forall x \in \mathcal{O}, \quad \varepsilon_0 > 0, \quad \forall \varepsilon \in [0,\varepsilon_0]$$

$$\{p \in \mathbb{R}^N, \quad H(x,u(x),p) \leq \varepsilon\} \text{ is convex.} \quad □$$

Proof: i) Let ω be a relatively compact open set properly contained in $\mathcal{O}$ ($\bar{\omega} \subset \mathcal{O}$) so that $u \in W^{1,\infty}(\omega)$ and $u_\varepsilon = u * p_\varepsilon$ is bounded in $W^{1,\infty}(\omega)$ - where $p_\varepsilon = \frac{1}{\varepsilon^N} p(\frac{\cdot}{\varepsilon})$, $p \in \mathcal{D}_+(\mathbb{R}^N)$, $\int p\,dx = 1$. The continuity of H yields that:

$$H(x,u_\varepsilon,Du_\varepsilon) \leq H(x,u,Du_\varepsilon) + \delta(\varepsilon) \text{ in } \bar{\omega}, \text{ with } \delta(\varepsilon) \to 0, \ \varepsilon \to 0$$

In addition by Jensen inequality:

$$\begin{aligned} H(x,u,Du_\varepsilon) &\leq \int H(x,u(x),Du(y))\, p_\varepsilon(x-y)dy \quad \text{in } \bar{\omega} \\ &\leq \mu(\varepsilon) \to 0 \quad \text{as } \varepsilon \to 0_+. \end{aligned}$$

Since $u_\varepsilon \in C^1$, u_ε is a viscosity subsolution and we may apply Proposition 3.

ii) In view of the semi-concavity of u, if $u-\phi$ has a local minimum at x and $\phi \in C^1(\mathcal{O})$, u is differentiable at x and Du(x) is limit of Du(y) as $y \to x$, $y \in \mathcal{O}$, y differentiability point of u. Since u is a.e. differentiable and the inequality holds a.e., we conclude.

II Existence, uniqueness and comparison results

We begin with the following model case:

(9) $H(x,u,Du) = 0$ in $\mathbb{R}^N$,

we will compare u with the solution v of

(10) $H(x,v,Dv) = f(x)$ in $\mathbb{R}^N$

and we will assume:

(11) $H(x,t,p) \in BUC(\mathbb{R}^N \times [-R,+R] \times \bar{B}_R)$ $\forall R < \infty$, $f \in C_b(\mathbb{R}^N)$;

(12) $\exists\ \lambda > 0$, $H(x,t,p) - H(x,s,p) \geq \lambda(t-s)$, $\forall x,p \in \mathbb{R}^N$, $\forall t \geq s \in \mathbb{R}$;

(13) $|H(x,t,p) - H(y,t,p)| \leq \omega_R(|x-y|(1+|p|))$ if $|t| \leq R$, $x,y,p, \in \mathbb{R}^N$

where $\omega_R(t) \to 0$ if $t \to 0_+$. We will also use the following condition (less general than (13)):

(14) $|H(x,t,p) - H(y,t,p)| \leq C_R^1|x-y|(1+|p|) + C_R^2|x-y|$
for $|t| \leq R$, $x,y,p \in \mathbb{R}^N$;

and

(15) $H(x,t,p) \to +\infty$ as $|p| \to \infty$, uniformly for $x \in \mathbb{R}^N$, t bounded.

<u>Theorem 5</u>: We assume (11), (12)

i) Let $u,v \in BUC(\mathbb{R}^N)$ be respectively viscosity sub and supersolution of (9) and (10). Assume either (13), or (15), then we have:

(16) $$\sup_{\mathbb{R}^N} (u-v)^+ \leq \frac{1}{\lambda} \sup_{\mathbb{R}^N} f^- .$$

ii) If (13), or (15) holds, there exists a unique $u \in BUC(\mathbb{R}^N)$ viscosity solution of (9).

iii) If (15) holds, $u \in W^{1,\infty}(\mathbb{R}^N)$; while if (14) holds and $\lambda_0 = C_R^1$ with $R = \|u\|_\infty$, then $u \in C^{0,\gamma}(\mathbb{R}^N)$ with $\gamma = \lambda/\lambda_0$ if $\lambda < \lambda_0$, $\gamma \in]0,1[$ if $\lambda = \lambda_0$, $\gamma = 1$ if $\lambda > \lambda_0$. □

<u>Remarks</u>: 1) Part i) (which implies uniqueness) is taken from M.G. Crandall and P.L. Lions [9]. Part iii) is taken from P.L. Lions [16], [17]. Part ii) if (15) or (14) hold is due to P.L. Lions [16], [17], see also P.E. Souganidis [24], while the general case of condition (13)

is due to G. Barles [1].

2) Extensions and variants of this result may be found in [9]. The case when H, f, u, v are not necessarily bounded is treated in H. Ishii [15], M.G. Crandall and P.L. Lions [10]. □

We consider now the case of boundary conditions: to simplify the presentation we consider a bounded open set $\mathcal{O}$ and we look for solutions of

$$(9') \qquad H(x,u,Du) = 0 \quad \text{in } \mathcal{O}$$

$$(10') \qquad H(x,v,Dv) = f \quad \text{in } \mathcal{O}.$$

Theorem 6: We assume (11), (12).

i) Let $u,v \in C(\bar{\mathcal{O}})$ be respectively viscosity sub and supersolution of (9'), (10'). Assume either (13), or (15) then we have:

$$(16') \qquad \max_{\mathcal{O}} (u-v)^+ \leq \max\{\frac{1}{\lambda} \max_{\bar{\mathcal{O}}} f^-, \max_{\partial\mathcal{O}} (u-v)^+\}.$$

ii) If (13) holds and if there exists $\underline{u},\bar{u} \in C(\mathcal{O})$ viscosity sub and super solutions of (9) with $\underline{u} = \bar{u}$ on $\partial\mathcal{O}$, there exists a unique $u \in C(\bar{\mathcal{O}})$ viscosity solution of (9) with $u = \underline{u} = \bar{u}$ on $\partial\mathcal{O}$.

iii) If (15) holds and if there exists $\underline{u} \in C(\bar{\mathcal{O}})$ viscosity subsolution of (9) there exists a unique viscosity solution $u \in C(\bar{\mathcal{O}})$ of (9) with $u = \underline{u}$ on $\partial\mathcal{O}$. In addition: $u \in W^{1,\infty}(\mathcal{O})$. □

Remarks: 1) Part i) is due to M.G. Crandall and P.L. Lions [9], part iii) is taken from P.L. Lions [16], [17]. In addition in [16], the existence of a subsolution is investigated if H is convex in (t,p). Part ii) was first proved in P.L. Lions [16], [17] in the special case when $\underline{u},\bar{u} \in W^{1,\infty}(\mathcal{O})$ and (14) holds, while the general case is proved in G. Barles [1], [2] by a very elegant method.

2) In general if $\lambda = 0$ (i.e. $H = H(x,p)$ for example), (16') and the uniqueness are not correct except if $H = H(x,p)$ satisfies:

$$H(x,p) \in C(\bar{\mathcal{O}} \times \mathbb{R}^N),\ H(x,p) \text{ is convex in } p \text{ for all } x \in \mathbb{R}^N$$
$$H(x,0) < 0 \quad \text{in } \bar{\mathcal{O}},\ (13) \text{ holds.}$$

(see [9] for more details).

3) The above results and remarks still hold if $\mathcal{O}$ is smooth and if we replace $\partial\mathcal{O}$ by a closed subset of $\partial\mathcal{O}$ Γ_+ such that for all $x_0 \in \Gamma_- = \partial\mathcal{O} - \Gamma_+$ we have uniformly for t,p bounded:

$$\underline{\lim}\ \{H(x,t,p+\lambda n(x))/\lambda \geq 0,\ \lambda\, d(x) \to 0,\ x \to x_0, x \in \mathcal{O}\} \geq H(x_0,t,p).$$

Here and below $d(x) = \mathrm{dist}(x,\partial\mathcal{O})$, $n(x)$ is the unit outward normal to $\partial\mathcal{O}$ at x if $x \in \partial\mathcal{O}$ and if x is near $\partial\mathcal{O}$, it is well known that there exists a unique $\bar{x} \in \partial\mathcal{O}$ such that $d(x) = |x - \bar{x}|$ and we set $n(x) = n(\bar{x})$ (observe that $n(x) = -\nabla d(x)$).

III Viscosity solutions and optimal deterministic control

We will consider here only a typical example of deterministic control problem: the case of infinite horizon problems with systems stopped at the first exit time of a domain defining the state constraints. Thus the state of the system is the solution X_t of

(17) $\dot{X}_t = b(X_t, \alpha_t)$ for $t \geq 0$, $X_0 = x \in \mathbb{R}^N$

where α_t - the control process - is a measurable function of t taking its values in A (given metric space) and where $b(x,\alpha) = b_\alpha(x)$ satisfy:

(18) $\sup_{\alpha \in A} \|b_\alpha\|_{W^{1,\infty}(\mathbb{R}^N)} < \infty.$

We then define a cost function:

(19) $J(x,\alpha_t) = \int_0^\tau f(X_t,\alpha_t)\, e^{-\lambda t} dt + \phi(X_\tau)\, e^{-\lambda\tau}$

where $\lambda > 0$, $\phi \in BUC(\mathbb{R}^N)$, $f(x,\alpha) = f_\alpha(x)$ satisfy:

(20) $\sup_{\alpha \in A} \|f_\alpha\|_{C_b(\mathbb{R}^N)} < \infty$, f_α is unif. cont. on $\mathbb{R}^N$, unif. in $\alpha \in A$.

To represent the constraints on X_t, we consider an open set $\mathcal{O}$ of

$\mathbb{R}^N$ and we will consider only the two cases when $\mathcal{O}$ is bounded, or when $\mathcal{O} = \mathbb{R}^N$ (no constraints). If $\mathcal{O}$ is bounded, we set $\tau = \inf\,(t \geq 0,\ X_t \notin \bar{\mathcal{O}})$ ($\tau = +\infty$ if $X_t \in \bar{\mathcal{O}}\ \ \forall t \geq 0$) - we might as well consider $\tau' = \inf\,(t \geq 0,\ X_t \notin \mathcal{O})$, or even any $\tilde{\tau}$ in $[\tau', \tau]$ -. If $\mathcal{O} = \mathbb{R}^N$, $\tau = +\infty$. In both cases in view of (20) and since $\lambda > 0$, J is well defined by (19). We finally define the value function $u(x) = \inf\limits_{\alpha_t} J(x, \alpha_t)$, $\forall x \in \bar{\mathcal{O}}$.

As it is well known (see for example W.H. Fleming and R. Rishel [13], P.L. Lions [16]) the dynamic programming principle applies to this control problem and yields:

i) the optimality principle: for any x, α_t, choose $T \in [0,\infty]$, then

$$u(x) = \inf_{\alpha_t}\{\int_0^{T\wedge\tau} f(X_t, \alpha_t) e^{-\lambda t} dt + u(X_T) e^{-\lambda T} 1_{(T<\tau)} + $$

$$+ \phi(X_\tau)\, e^{-\lambda\tau}\, 1_{(\tau \leq T)}\}.$$

ii) the "Bellman equation": assume u is differentiable at $x_0 \in \mathcal{O}$ then "differentiating the above equality" with respect to T, one finds:

$$\text{(B)}\quad \sup_{\alpha \in A}\ [-b_\alpha(x_0)\,.\nabla u(x_0) + \lambda u(x_0) - f_\alpha(x_0)] = 0\,.$$

Of course (B) is a special case of (HJ) for the Hamiltonian:

$$H(x,t,p) = \sup_{\alpha \in A}\ [-b_\alpha(x).p - f_\alpha(x)] + \lambda t$$

and H satisfies (11), (12), (13). In addition H is convex in p (in (t,p)).

The following result (taken from P.L. Lions [16]) is an illustration of the following general idea: value functions are always viscosity solutions (provided they are continuous!).

<u>Proposition 7</u>: We assume (18), (20).

i) If $\mathcal{O} = \mathbb{R}^N$, $u \in BUC(\mathbb{R}^N)$ is the unique viscosity solution of (B) in $BUC(\mathbb{R}^N)$.

ii) If $\mathcal{O} \neq \mathbb{R}^N$ and if $u \in C(\mathcal{O})$, then u is a viscosity solution of (B). And if $u \in C(\bar{\mathcal{O}})$, u is the unique viscosity solution v of (B) in $C(\bar{\mathcal{O}})$ satisfying $v = u$ on $\partial\mathcal{O}$. □

Remarks: 1) With the notations of Remark 3) following Theorem 6, we see that u is the unique viscosity solution v of (B) in $C(\bar{\mathcal{O}})$ satisfying v = u on Γ_+, and very often one has $u = \phi$ on Γ_+.
2) The same proof as in part A) of Proposition 4 shows that if $v \in C(\mathcal{O})$ satisfies:

$$\forall \alpha \in A, \; - b_\alpha(x).\nabla v + \lambda v - f_\alpha(x) \leq 0 \text{ in } \mathcal{D}'(\mathcal{O})$$

then v is a viscosity subsolution of (B). And using Theorem 6, we see that if $u, v \in C(\bar{\mathcal{O}})$ and $v \leq u$ on $\partial\mathcal{O}$, then $v \leq u$ on $\mathcal{O}$. Thus we find back the fact that u is the maximal subsolution of (B) (even in very weak senses). □

Proof: Let $x_0 \in \mathcal{O}$ be a global minimum point (for example) of $u-\phi$ where $\phi \in C^1(\mathcal{O})$. Since u is bounded, we may consider $\phi \in C^1_b(\mathcal{O})$. We have to check (8) at x_0: we first observe that in both cases ($\mathcal{O} = \mathbb{R}^N$ or $\mathcal{O} \neq \mathbb{R}^N$) for T small independent of α_t

$$u(x_0) = \inf_{\alpha_t} \{\int_0^T f(X_t, \alpha_t)e^{-\lambda t}dt + u(X_T)\, e^{-\lambda T}\} .$$

Indeed: $|X_t - x_0| \leq Ct$ and thus $\tau \geq \dfrac{\text{dist}(x_0, \partial\mathcal{O})}{C}$.

Since only the gradient of ϕ at x_0 occurs in (8), ϕ is in fact defined up to a constant and we may choose without loss of generality

$$\phi(x_0) = u(x_0) \; , \quad \text{thus} \quad u \geq \phi \text{ in } \mathcal{O} .$$

Therefore we obtain for all T small:

$$\phi(x_0) \geq \inf_{\alpha_t} \{\int_0^T f(X_t, \alpha_t)e^{-\lambda t}dt + \phi(X_T)\, e^{-\lambda T}\}$$

and the usual derivation of (B) then applies giving:

$$\sup_{\alpha \in A} [- b_\alpha(x_0).\nabla\phi(x_0) + \lambda\phi(x_0) - f_\alpha(x_0)] \geq 0$$

i.e. (8) ! □

The converse statement is given by the following result (taken from P.L. Lions [18], [19], P.L. Lions and P.E. Souganidis [23]): we need a few notations, for $\delta > 0$, we set $\mathcal{O}_\delta = \{x \in \mathcal{O}, \mathrm{dist}(x, \partial\mathcal{O}) > \delta\}$ and $\tau_\delta = \inf(t \geq 0, X_t \in \overline{\mathcal{O}}_\delta)$.

<u>Theorem 8</u>: Let $v \in C(\overline{\mathcal{O}})$.

i) v is a viscosity subsolution of (B) if and only if we have for all $\delta > 0$

$$v(x) \leq \inf_{\alpha_t}\{\int_0^{T\wedge\tau_\delta} f(X_t, \alpha_t) e^{-\lambda t} dt + v(X_{T\wedge\tau_\delta}) e^{-\lambda T\wedge\tau_\delta}, \forall x \in \mathcal{O}$$

where $T \in [0,\infty]$ may depend on x, α_t, δ.

ii) v is a viscosity supersolution of (B) if and only if we have for all $\delta > 0$

$$v(x) \geq \inf_{\alpha_t}\{\int_0^{T\wedge\tau} f(X_t, \alpha_t) e^{-\lambda t} dt + v(X_{T\wedge\tau_\delta}) e^{-\lambda T\wedge\tau_\delta}\}, \quad \forall x \in \mathcal{O}$$

where $T \in [0,\infty]$ may depend on x, α_t, δ. □

If v is locally Lipschitz, a further formulation of (B) exists: we introduce for any $x \in \mathcal{O}$, the set

$$K_x = \overline{\mathrm{co}}(b_\alpha(x), f_\alpha(x)) / \alpha \in A\}$$

<u>Theorem 9</u>: Let $v \in W^{1,\infty}_{loc}(\mathcal{O})$.

i) v is a viscosity subsolution of (B) if and only if

$$(21) \quad \sup_{(b,f)\in K_x} \{\overline{\lim_{t\to 0_+}} \frac{1}{t}[u(x) - u(x+tb)] - f\} + \lambda u(x) \leq 0, \ \forall x \in \mathcal{O}$$

ii) v is a viscosity supersolution of (B) if and only if

$$(22) \quad \sup_{(b,f)\in K_x} \{\overline{\lim_{t\to 0_+}} \frac{1}{t}[u(x) - u(x+tb)] - f\} + \lambda u(x) \geq 0, \ \forall x \in \mathcal{O}$$

□

<u>Remarks</u>: 1) This result and extensions to differential games may be found in [11], [23]. If $u \in C(\mathcal{O})$ satisfies (21) (resp. (22)), u is a viscosity subsolution (resp. supersolution).

2) Of course v is a viscosity solution of (B) if and only if

23) $\sup\limits_{(b,f)\in K_x} \{\overline{\lim\limits_{t\to 0_+}} \frac{1}{t}[u(x)-u(x+tb)] - f\} + \lambda u(x) = 0 \ , \ \forall x \in \mathcal{O}$

3) This "everywhere" formulation of (B) has the advantage of implying obvious uniqueness and comparison results: indeed if $u,v \in C(\overline{\mathcal{O}})$ satisfy (21), (22) respectively, we know by Theorem 6 that $\max\limits_{\overline{\mathcal{O}}} (u - v)^+ = \max\limits_{\partial\mathcal{O}} (u - v)^+$. This may be seen directly using (22), (21) since if

$$(u - v)(x_o) = \max_{\overline{\mathcal{O}}} (u - v)^+ > \max_{\partial\mathcal{O}} (u - v)^+ \ ,$$

we have clearly for any $(b,f)\in K_x$

$$\overline{\lim_{t\to 0_+}} \frac{1}{t}[\, u(x_o) - u(x_o + tb)\,] - f \geq \overline{\lim_{t\to 0_+}} \frac{1}{t}[v(x_o) - v(x_o+ tb)] - f$$

and in view of (21), (22) we deduce

$$\lambda u(x_o) \leq \lambda v(x_o)$$

and the contradiction proves our claim. □

The fact that v being a viscosity sub or supersolution of (B) satisfies (21) or (22) is a straightforward consequence of Theorem 8 and follows from a careful examination of the standard derivation of (B). The converse is even easier since if $v \in C(\mathcal{O})$ satisfies (22) and $v - \phi$ admits at a point $x_o\in \mathcal{O}$ a local minimum we have easily

$$\sup_{(b,f)\in K_x} \{\overline{\lim_{t\to 0_+}} \frac{1}{t}[\phi(x_o)-\phi(x_o+tb)]- f\} \geq \sup_{(b,f)\in K_x} \{\overline{\lim_{t\to 0_+}} \frac{1}{t}[u(x_o)-u(x_o+tb)]\}$$

□

IV State-constraints problems

In many applications the state-constraints are formulated in the following way: a control α_t is said to be admissible (we will denote this by $\alpha_t\in A_x$) if the state process X given by (17) (for $x \in\overline{\mathcal{O}}$) satisfies

(24) $\quad X_t \in \bar{\mathcal{O}}, \quad \forall t \geq 0$.

The value function is now defined by

$$u(x) = \inf_{\alpha_t \in A_x} J(x,\alpha_t) \quad (u(x)=+\infty \text{ if } A_x=\emptyset), \forall x \in \mathcal{O}.$$

From now on, to simplify, we assume that $\mathcal{O}$ is a bounded, smooth open set of $\mathbb{R}^N$.

The question of finding "good" characterizations of u is mainly open in full generality: we would like to present here two cases where we are able to solve this question.

First of all we assume

(25) $\quad \forall (x,\alpha) \in \partial\mathcal{O} \times A, \ (b(x,\alpha),n(x)) \leq 0$

This easily yields that all controls α_t are admissible for all $x \in \mathcal{O}$ and $\Gamma_- = \partial\mathcal{O}$, $\Gamma_+ = \emptyset$. In this case we have immediately by the results and remarks of the preceding sections:

i) $u \in C(\bar{\mathcal{O}})$ is the unique viscosity solution of (B)

ii) u is the maximum element of the set of subsolutions of (B) (either in viscosity sense, or in distributions sense).

The second case we can treat is when:

(26) $\quad \exists \delta > 0, \ \forall x \in \partial\mathcal{O}, \ \bar{B}_\delta \subset \overline{\text{co}}\{b(x,\alpha)/\alpha \in A\}$.

Then simple considerations yield that $u \in C(\mathcal{O})$, u is Lipschitz in a neighbourhood of $\partial\mathcal{O}$. Again one shows easily that u is the maximum subsolution of (B) (in viscosity sense, in distributions sense...) and that u is the limit, as ε goes to 0, in $C(\mathcal{O})$ of the viscosity solution $u_\varepsilon \in B\,UC(\mathbb{R}^N)$ of the following penalized problem:

(27) $$\sup_{\alpha \in A} [-b_\alpha(x).Du_\varepsilon + \lambda u_\varepsilon - f_\alpha(x)] = \frac{1}{\varepsilon} p \text{ in } \mathbb{R}^N$$

where $\varepsilon > 0$, $p \in B\,UC(\mathbb{R}^N)$, $p \equiv 0$ in $\mathcal{O}$, $p \geq \mu(\delta) > 0$ if $\text{dist}(x,\bar{\mathcal{O}}) \geq \delta > 0$.

Furthermore, the dynamic programming principle implies as in Proposition 7 that u is a viscosity solution of (B): the only remaining question is to determine the boundary condition on u. To this end we introduce for $x \in \partial\mathcal{O}$ the following convex sets

$$K^\circ_x = \overline{co}\{(b_\alpha(x), f_\alpha(x)/\alpha \in A,\ (b_\alpha(x), n(x)) < 0\}$$

$$K_x = \overline{co}\{(b_\alpha(x), f_\alpha(x))/\alpha \in A\}$$

$$\tilde{K}^\circ_x = \{(b,f) \in K_x/(b, n(x)) \leq 0\}.$$

A proof similar to the proof of Theorem 9 (recall that u is Lipschitz near $\partial\mathcal{O}$) then yields:

$$(28)\qquad \sup_{(b,f)\in K^\circ_x} \{\overline{\lim_{t\to 0_+}} \tfrac{1}{t}[u(x)-u(x+tb)] - f\} + \lambda u(x) \leq 0$$

$$(29)\qquad \sup_{(b,f)\in \tilde{K}^\circ_x} \{\overline{\lim_{t\to 0_+}} \tfrac{1}{t}[u(x)-u(x+tb)] - f\} + \lambda u(x) \geq 0$$

We need to explain the meaning of $u(x+tb)$ in (29) since b only satisfies $(b,n(x))\leq 0$ and thus $x+tb$ may belong to $\mathbb{R}^N - \mathcal{O}$ (where u is not a priori defined), the precise meaning of $u(x+tb)$ is obtained as follows: we still denote by u any Lipschitz extension of u nearby $\partial\mathcal{O}$, then the lim does not depend on the Lipschitz extension of u. Indeed since $(b,n(x))\leq 0$, $\frac{1}{t}\,\mathrm{dist}(x+tb, \bar{\mathcal{O}}) \to 0$ as $t \to 0_+$.

Hence, if we assume:

$$(30)\qquad \tilde{K}^\circ_x = K^\circ_x,\ \forall x \in \partial\mathcal{O},$$

we have obviously:

$$(31)\qquad \sup_{(b,f)\in K^\circ_x} \{\overline{\lim_{t\to 0_+}} \tfrac{1}{t}[u(x)-u(x+tb)]-f\} + \lambda u(x) = 0,\quad \forall x \in \partial\mathcal{O}$$

We may now state the:

<u>Theorem 10</u>: We assume (26). Then $u \in C(\bar{\mathcal{O}})$ and u is Lipschitz near $\partial\mathcal{O}$. If we assume in addition (30), u is the unique viscosity solution of (B) in $C(\bar{\mathcal{O}})$ satisfying the boundary condition (31). □

<u>Remark</u>: Observe that (31) is essentially the Bellman equation on $\partial\mathcal{O}$

restricted to (relaxed) controls such that the corresponding vector field points inward. □

There just remains to prove the uniqueness of u: indeed if v is another viscosity solution of (B), we know by Theorem 6 that

$$\max_{\bar{\mathcal{O}}} |u - v| = \max_{\partial\mathcal{O}} |u - v|$$

and if for example $\max_{\partial\mathcal{O}} |u - v| = (u - v)(x_0)$, $x_0 \in \partial\mathcal{O}$; then the same proof as in Remark 3) following Theorem 9 gives

$$(u - v)(x_0) \leq 0$$

and thus $u \equiv v$. □

V Extension to second order equations

We now want to indicate how one can extend the previous considerations to fully nonlinear, degenerate elliptic, second-order equations:

(32) $\quad H(x,u,Du,D^2u) = 0 \quad \text{in } \mathcal{O}$

where H is continuous on $\mathcal{O} \times \mathbb{R} \times \mathbb{R}^N \times Y^N$ – we denote by Y^N the space of $N \times N$ symmetric matrices – and H satisfies:

(33) $\quad H(x,t,p,M_1) \geq H(x,t,p,M_2) \quad \text{if } M_1 \leq M_2$.

The natural extension of viscosity solutions will use the following notion of superdifferential of order 2 (denoted by $D_2^+ v(x)$) of at the point $x \in \mathcal{O}$:

$$D_2^+v(x) = \{(\xi,M) \in \mathbb{R}^N \times Y^N /$$

$$\limsup_{y\to x, y\in\mathcal{O}} \{v(y)-v(x)-(\xi,y-x)-\tfrac{1}{2}(M(y-x),y-x)\}\,|y-x|^{-2} \leq 0\}$$

$$= \{(\xi,M) \in \mathbb{R}^N \times Y^N / \exists \phi \in C^2(\mathcal{O}), \phi(x) = v(x), D\phi(x) = \xi,$$

$$D^2\phi(x) = M \ , \quad \phi(y) > v(y) \text{ if } y \in \mathcal{O} \ , \ y \neq x \} \ ,$$

and one defines similarly $D_2^- v(x)$ - the subdifferential of order 2 of v at x.

Clearly $D_2^+(v)$ is a closed convex set (possibly empty) and if $(\xi,M) \in D_2^+ v(x)$ then $(\xi,M') \in D_2^+ v(x)$ for all $M' \in Y^N$, $M' \geq M$.

<u>Remark</u>: If v is twice differentiable at x, then $D_2^+ v(x) = \{(Dv(x),M)/M \in Y^N, M \geq D^2 v(x)\}$, and $D_2^- v(x) = \{(Dv(x),M)/M \in Y^N, M \leq D^2 v(x)\}$. □

<u>Definition</u>: Let $u \in C(\mathcal{O})$; u is said to be a viscosity subsolution (resp. supersolution) of (32) if

(34) $H(x,u(x),\xi,M) \leq 0 \ , \ \forall x \in \mathcal{O}, \ \ \forall (\xi,M) \in D_2^+ u(x)$;

(resp.

(35) $H(x,u(x),\xi,M) \geq 0 \ , \ \forall x \in \mathcal{O}, \ \ \forall (\xi,M) \in D_2^- u(x)$).

And u is a viscosity solution of (32) is u is a viscosity sub and super solution. □

With this notion, the results mentioned in section I are easily adapted (see P.L. Lions [18], [21] where additional properties of this notion are discussed). The major open question in this field concerns the uniqueness results of section II (existence of such weak solutions is not difficult under convenient structure conditions on H). Such uniqueness results would be determinant in order to have a satisfactory theory of stochastic differential games.

However the results of section III may be extended to the second-order case and to the Hamilton-Jacobi-Bellman equations associated with optimal stochastic control:

(HJB) $\sup_{\alpha \in A} \ [A_\alpha u - f_\alpha(x)] = 0 \text{ in } \mathcal{O}$

where $(A_\alpha)_{\alpha \in A}$ is a family of second order, elliptic, degenerate operators with smooth coefficients. It is proved in [18], [19] that the analogues of Proposition 7 and Theorem 8 are still valid in this situation; the proof using heavily the convexity in (p,M) of the nonlinearity H occurring in (HJB) equations (which implies the existence of smooth subsolutions of (HJB) by arguments related to the proof of part i) of Proposition 4).

Bibliography:

[1] G. Barles: Existence results for first order Hamilton-Jacobi equations. To appear in Ann. I.H.P. Anal. Non Lin., see also Thèse de 3e. cycle, Paris-Dauphine, 1983.

[2] G. Barles: Some remarks on existence results for first-order Hamilton-Jacobi equations, Submitted to Ann. I.H.P. Anal. Non Lin.

[3] G. Barles: Quasi-variational inequalities and first-order Hamilton-Jacobi equations. Thèse de 3e. cycle, Paris-Dauphine, 1983.

[4] G. Barles: Determinisitc impulsive control problems. To appear in SIAM J. Control. Optim., see also Thèse de 3e. cycle, Paris-Dauphine, 1983.

[5] I. Capuzzo-Dolcetta: On a discrete approximation of the Hamilton-Jacobi equation of dynamic programming. Appl. Math. Optim. 10 (1983), p. 367-377.

[6] I. Capuzzo-Dolcetta and L.C. Evans: Optimal switching for ordinary differential equations. SIAM J. Control Optim., 22 (1984) p.143-46.

[7] M.G. Crandall, L.C. Evans and P.L. Lions: Some properties of viscosity solutions of Hamilton-Jacobi equations. Trans. Amer. Math. Soc., 1984.

[8] M.G. Crandall and P.L. Lions: Conditions d'unicité pour les solutions généralisées des equations de Hamilton-Jacobi du premier ordre. C.R. Acad. Sci. Paris, 292 (1981), p. 183-186.

[9] M.G. Crandall and P.L. Lions: Viscosity solutions of Hamilton-Jacobi equations. Trans Amer. Math. Soc., 277 (1983), p. 1-42.

[10] M.G. Crandall and P.L. Lions: Solutions de viscosité non barnées des equations de Hamilton-Jacobi du premier ordre. C.R. Acad. Sci. Paris, 1984.

[11] L.C. Evans and H. Ishii: Differential games and nonlinear first order PDE on bounded domains, to appear.

[12] L.C. Evans and H. Ishii: A PDE approach to some asymptotic problems concerning random differential equations with small noise intensities, to appear.

[13] W.H. Fleming and R. Rishel: Deterministic and stochastic optimal control. Springer, Berlin, 1975.

[14] W.H. Fleming and P.E. Souganidis: to appear.

[15] H. Ishii: Uniqueness of unbounded viscosity solutions of Hamilton-Jacobi equations. To appear in Ind. Univ. Math. J.

[16] P.L. Lions: Generalized solutions of Hamilton-Jacobi equations. Pitman, London, 1982.

[17] P.L. Lions: Existence results for first-order Hamilton-Jacobi equations. Ricerche di Mat., 32 (1983), p. 3-23

[18] P.L. Lions: Optimal control of diffusion processes and Hamilton-Jacobi-Bellman equations. Parts 1,2. Comm. P.D.E., 8 (1983), p. 1101-1174, p. 1229-1276.

[19] P.L. Lions: Some recent results in the optimal control of diffusion processes. Stochastic Analysis, Proceedings of the Taniguchi International Symposium on Stochastic Analysis, Katata and Kyoto, 1982. Kinokuniya, Tokyo, 1984.

[20] P.L. Lions: On the Hamilton-Jacobi-Bellman equations. Acta Applicanda, 1 (1983), p. 17-41.

[21] P.L. Lions: Fully nonlinear elliptic equations and applications. In Nonlinear Analysis, Function Spaces and Applications, Teubner, Leipzig, 1982.

[22] P.L. Lions and M. Nisio: A uniqueness result for the semigroup associated with the Hamilton-Jacobi-Bellman operator. Proc. Japan Acad., 58 (1982), p. 273-276.

[23] P.L. Lions and P.E. Souganidis: Differential games, optimal control and directional derivatives of viscosity solutions of Bellman's and Isaacs' equations. To appear in SIAM J. Control Optim.

[24] P.E. Souganidis: PhD Thesis, University of Wisconsin-Madison, 1983.

SOME CONTROL PROBLEMS OF DEGENERATE DIFFUSIONS WITH UNBOUNDED COST

José Luis MENALDI (*)
Department of Mathematics
Wayne State University
Detroit, Michigan 48202
U.S.A.

and

Maurice ROBIN
INRIA - Rocquencourt
Domaine de Voluceau
B.P.105
78153 Le Chesnay Cedex
France

ABSTRACT

A dynamic programming approach is used for a class of optimal control problems for diffusion processes with jumps. The control of the system is an adapted process with bounded variation, which acts continuously and impulsively on the system. This class of problems includes for instance, the so-called cheap control problems and monotone follower problems. Results concerning the characterization of the optimal cost and the construction of an optimal feedback law are established.

INTRODUCTION

The purpose of this paper is to present some stochastic control of degenerate diffusion processes with unbounded cost in which the optimal control is expected to have bounded variation and to be not absolutely continuous. This class of control affects the dynamic of the system in an additive manner, even it may cause some discontinuities on the state of the system.

This kind of problems are motivated by the study of the asymptotic behaviour of the classical impulse control problems as the fixed cost of impulse tends to zero. The same situation occurs when we let the unit cost per control go to zero in a standard stochastic control problem. The typical case is the so-called cheap control problems and monotone follower problems.

For instance, let the state of the dynamic system be described as

$$y(t,\nu) = x + w_t + N_t + \nu(t) \quad , \quad t \geq 0 \; , \tag{1}$$

(*) This research has been supported in part by U.S. Army Research Office Contract DAAG29-83-K-0014 and completed during a visit at the INRIA and the University Paris-Orsay.

where $(w_t, t \geq 0)$ and $(N_t, t \geq 0)$ are standard one-dimensional Wiener and Poisson processes, respectively, and $(\nu(t), t \geq 0)$ is the control. In general $\nu(t)$ represents the resources spent in the system up to the time $t \geq 0$. If the system reacts almost instantaneously with respect to the resources used, it is plausible to allow possible jumps for $\nu(t)$. However the variation process

$$\eta(t) = |\nu(0)| + \sup\{\sum_{i=1}^{n} |\nu(t_i) - \nu(t_{i-1})| : 0=t_0<t_1<\ldots<t_n=t\} \quad (2)$$

needs to be finite at each time $t \geq 0$.

Consider the problem of minimizing

$$J_x(\nu) = E\{\int_0^\infty [|y(t,\nu)|^2 + \alpha\, c\, \eta(t)]\, e^{-\alpha t}\, dt\}, \quad (3)$$

for a given discount factor $\alpha > 0$ and some constant $c > 0$. Note that if ν is absolutely continuous, then

$$\eta(T) = |\nu(0)| + \int_0^T |\dot{\nu}(t)|dt. \quad (4)$$

A formal application of the dynamic programming arguments to the optimal cost

$$u(x) = \inf\{J_x(\nu) : \nu\} \quad (5)$$

yields the Hamilton-Jacobi-Bellman equation

$$(Au - f) \wedge (c - |\dot{u}|) = 0 \text{ in } \mathbb{R}, \quad (6)$$

where the dot means derivative, $f(x) = x^2$ and

$$\left.\begin{aligned} &Au(x) = -\frac{1}{2}\ddot{u}(x) - \lambda(u(x+1) - u(x)) + \alpha u(x), \\ &E\{N_t\} = \lambda t \quad, \quad t \geq 0 \quad, \quad \lambda \geq 0. \end{aligned}\right\} \quad (7)$$

This is a limit case of a classical "continuous control" problem, namely : $\varepsilon \to 0$,

$$\left.\begin{aligned} u_\varepsilon(x) &= \inf\{\, J_x^\varepsilon(\dot\nu) \;:\; \dot\nu\},\\ J_x^\varepsilon(\dot\nu) &= E\{\int_0^\infty [|y(t,\dot\nu)|^2 + c\,|\dot\nu(t)|]e^{-\alpha t}\,dt + \varepsilon\int_0^\infty |\dot\nu(t)|^2 e^{-\alpha t}\,dt\}\\ y(t,\dot\nu) &= x + w_t + N_t + \int_0^t \dot\nu(s)ds. \end{aligned}\right\} \quad (8)$$

On the other hand, it is a limit case of a classical "impulse control", namely : $\varepsilon \to 0$,

$$\left.\begin{aligned} u_\varepsilon(x) &= \inf\,\{J_x^\varepsilon(\nu) \;:\; \nu = (\theta_i,\ \xi_i,\ i= 1,2,\ldots)\}\ ,\\ J_x^\varepsilon(\nu) &= E\,\{\int_0^\infty |y(t,\nu)|^2\, e^{-\alpha t}\,dt + \sum_{i=1}^\infty (\varepsilon + c|\xi_i|)e^{-\alpha\theta_i}\}\ ,\\ y(t,\nu) &\text{ given by (1) with } \nu(t) = \sum_{i=1}^\infty \xi_i 1(\theta_i \le t). \end{aligned}\right\} \quad (9)$$

Several modifications are permited in this preliminary example. In particular, we may assume that

$$\nu(t) \ge \nu(s) \ge 0 \quad , \quad \forall t \ge s \ge 0. \qquad (10)$$

Then, the equation (6) becomes

$$(Au - f) \wedge (\dot u + c) = 0 \text{ in } \mathbb{R} \qquad (11)$$

and the case $c = 0$ is now meaningful. This makes clear the terminology of "monotone follower " and "cheap control" problems.
As main references for similar problems we can mention the works of Barron and Jensen [1] , Bather and Chernoff [2], Benes et al.[3], Borodovskii et al.[6], Bratus [7], Chernousko [8,9], Francis and Glover [12], Gorbunov [14], Harrison and Taylor [15], Jameson and O'Malley [16] , Karatzas [17,18] , Kokotovic et al.[19], Menaldi and Robin [22], and several other recent papers.

In this article, we review and extend some of the results given in Chow et al. [10] and [22,23]. The plan is the following :

1. Statement of the Problem.
2. A Nonlinear Semigroup.
3. Variational Formulation.
4. One Dimensional Case.

Troughout this paper, the statement of theorems are precise, but only the main ideas of the proof are given.

1. STATEMENT OF THE PROBLEM

Let $(\Omega, F, P, F_t, w_t, \mu_t, t \geq 0)$ be a complete Wiener-Poisson space in $\mathbb{R}^n \times \mathbb{R}^m_*$, with Levy measure π on $\mathbb{R}^m_* = \mathbb{R}^m - \{0\}$, i.e. (Ω,F,P) is a complete probability space, $(F_t, t \geq 0)$ is an increasing right continuous family of complete sub σ-algebras of F, $(w_t, t \geq 0)$ is a standard Wiener-Process in $\mathbb{R}^n$ w.r.t. $(F_t, t \geq 0)$, and $(\mu_t, t \geq 0)$ is a standard Poisson martingale measure in $\mathbb{R}^m_*$ w.r.t. $(F_t, t \geq 0)$ and independent of $(w_t, t \geq 0)$. The measure π is related to $(\mu_t, t \geq 0)$ as follows : for every compact subset A of $\mathbb{R}^m_*$, the process $(\mu_t(A))^2 - t\,\pi(A)$ is a martingale w.r.t.$(F_t, t \geq 0)$. Starting with a random Poisson measure $p(t,A)$ we may define μ_t, π by means of relations $\mu_t(A) = p(t,A) - E\{p(t,A)\}$, $t\,\pi(A) = E\{p(t,A)\}$. For more details see Bensoussan and Lions [5], Gihman and Skorodhod [13].

Suppose we are given measurable functions $g = (g_i(x), i=1,\ldots d)$, $\sigma = (\sigma_{ik}(x), i=1,\ldots,d, k=1,\ldots,n)$, $\gamma = (\gamma_i(x,\zeta), i=1,\ldots,d)$, x in $\mathbb{R}^d$, ζ in $\mathbb{R}^m_*$ such that for every $p \geq 2$ there exists a constant C_p satisfying :

$$|g(x)|^p + |\sigma(x)|^p + \int_{\mathbb{R}^m_*} |\gamma(x,\zeta)|^p \,\pi(d\zeta) \leq C_p(1+|x|^p) \tag{1.1}$$

$$\left.\begin{aligned} &|g(x)-g(x')|^p + |\sigma(x)-\sigma(x')|^p + \int_{\mathbb{R}^m_*} |\gamma(x,\zeta)-\gamma(x',\zeta)|^p \pi(d\zeta) \leq \\ &\qquad \leq C_p|x-x'|^p, \end{aligned}\right\} \tag{1.2}$$

for any x, x' in $\mathbb{R}^d$ and where $|.|$ denotes the appropriate Euclidean norm.

The uncontrolled evolution of the dynamical system is a diffusion process with jumps whose coefficients have been identified to be $g(x)$, $\sigma(x)$, $\gamma(x,\zeta)$. This means that if $y^o(t) = y^o_x(t,\omega)$, $\omega \in \Omega$, represents the state of the system at the time t we have

$$\left.\begin{aligned} &dy^o(t) = g(y^o(t))dt + \sigma(y^o(t))dw_t + \int_{\mathbb{R}^m_*} \gamma(y^o(t),\zeta)d\mu_t(\zeta), \quad t \geq 0, \\ &y^o(0) = x \end{aligned}\right\} \tag{1.3}$$

where x is the initial state. Note that either $\sigma = 0$ or $\gamma = 0$ are permitted.

The control is to modify the state of the system by adding a new stochastic process which has locally bounded variation. The cost of such an intervention has also two terms, one related to the evolution of the state and another depending on the variation of the control.

Then an additive control is a stochastic process $(\nu(t),\ t \geq 0)$ which is right continuous having left-hand limits, progressively measurable w.r.t. $(F_t,\ t \geq 0)$. Also, it has bounded variation on each compact set $[0,T]$, $T > 0$ and takes values in some subset of $\mathbb{R}^d$, for instance $\nu(t) \geq 0$, for every $t \geq 0$. The controlled state follows the stochastic equations

$$\left.\begin{aligned} y(t) = x + \nu(t) + \int_0^t g(y(s))dt + \int_0^t \sigma(y(s))dw_s + \\ \int_0^t \int_{\mathbb{R}^m_*} \gamma(y(s),\zeta)d\mu_s(\zeta) \quad , \quad t \geq 0 \ , \end{aligned}\right\} \tag{1.4}$$

and the cost of an additive cost $(\nu(t),\ t \geq 0)$ is

$$\left.\begin{aligned} J_x(\nu) = E\{ \int_0^\infty f(y(t))\exp(-\alpha t)dt + c(\eta(0)) + \\ + \int_0^\infty \exp(-\alpha t)dc(\eta(t)) \quad , \end{aligned}\right\} \tag{1.5}$$

where α is a positive constant, f, c are given function satisfying suitable conditions and $\eta(t)$ denotes the variation of ν on $[0,t]$, i.e. $\eta = (\eta_i,\ i=1,\dots,d)$

$$\eta_i(t) = |\eta_i(0)| + \sup\{ \sum_{j=1}^n |\nu_i(t_j)-\nu_i(t_{j-1})| : 0=t_0<t_1<\dots<t_n=t\}. \tag{1.6}$$

The purpose is to characterize the optimal cost function

$$\hat{u}(x) = \inf\{ J_x(\nu) : \nu \} \tag{1.7}$$

and to obtain a feedback law which provides an optimal additive control $\hat{\nu}$.

A common assumption on f, c and α is the following : $p \geq 0$,

$$\left.\begin{aligned} & 0 \leq f(x) \ , \ c(x) \leq C(1 + |x|^p) \quad , \quad \forall x \in \mathbb{R}^d, \\ & \alpha - \beta(p,\lambda) \geq 2\alpha_0 > 0, \\ & f,\ c \text{ are lower semicontinuous} \end{aligned}\right\} \tag{1.8}$$

where C, λ are positive constants. The function $\beta(p,\lambda)$ is defined by

$$\left.\begin{aligned}\beta(p,\lambda) = \sup\{\, & p\, b_1(x)(\lambda+|x|^2)^{-1} + p(p-2)b_2(x)(\lambda+|x|^2)^{-2} + \\ & + b_3(x,p,\lambda)(\lambda+|x|^2)^{-p} : x \text{ in } \mathbb{R}^d \},\end{aligned}\right\} \quad (1.9)$$

with

$$b_1(x) = \sum_{i=1}^{d} x_i g_i(x) + \frac{1}{2} \sum_{i=1}^{d} \sum_{k=1}^{n} [\sigma_{ik}(x)]^2 ,$$

$$b_2(x) = \frac{1}{2} \sum_{i,j=1}^{d} x_i x_j \sum_{k=1}^{n} \sigma_{ik}(x)\sigma_{jk}(x),$$

$$b_3(x,p,\lambda) = \int_{\mathbb{R}^m_*} [(\lambda + |x + \gamma(w,\zeta)|^2)^{p/2} - (\gamma + |x|^2)^{p/2} - \\ -p(\lambda+|x|^2)^{p/2-1} \sum_{i=1}^{d} x_i \gamma_i(x,\zeta)]\pi(d\zeta).$$

Note that $\beta(p,\lambda)$ is finite under the assumption (1.1). Moreover, if we suppose that

$$|g(x)|^p + |\sigma(x)|^p + \int_{\mathbb{R}^m} |\gamma(x,\zeta)|^p \,\pi(d\zeta) \le C(1+|x|^{p-\varepsilon}), \quad (1.10)$$

holds for every x in $\mathbb{R}^d$, $p \ge 2$ and some positive constants $C=C(p), \varepsilon=\varepsilon(p)$, then for any $p > 0$

$$\beta(p,\lambda) \to 0 \text{ as } \lambda \to \infty \quad . \quad (1.11)$$

By convention, we set $\beta(p,\lambda) = 0$ for $p = 0$. In this case $f(x)$ and $c(x)$ are bounded.

Notice that for every $t \ge 0$, x in $\mathbb{R}^d$ we have

$$E\,\{(\lambda + |y_x^o(t)|^2)^{p/2} \exp(-\beta(p,\lambda)t\} \le (\lambda+|x|^2)^{p/2}. \quad (1.12)$$

Similarly, defining for $p > 0$, $\lambda > 0$,

$$\left.\begin{aligned}\beta'(p,\lambda) = \sup\{\, & p\, c_1(x+\xi,x'+\xi)(\lambda+|x-x'|^2)^{-1} + \\ & + p(p-2)c_2(x+\xi,x'+\xi)(\lambda+|x-x'|^2)^{-2} + \\ & + c_3(x+\xi,x'+\xi,p,\lambda)(\lambda+|x-x'|^2)^{-p} : x,x',\xi \text{ in } \mathbb{R}^d\},\end{aligned}\right\} \quad (1.13)$$

where

$$c_1(x,x') = \sum_{i=1}^{d}(x_i - x_i')(g_i(x)-g_i(x')) + $$
$$+ \frac{1}{2} \sum_{i,j=1}^{d} \sum_{k=1}^{n} (\sigma_{ik}(x)-\sigma_{ik}(x'))(\sigma_{jk}(x)-\sigma_{jk}(x')),$$

$$c_2(x,x') = \frac{1}{2} \sum_{i,j=1}^{d} (x_i-x_i')(x_j-x_j') \sum_{k=1}^{n} (\sigma_{ik}(x)-\sigma_{ik}(x'))(\sigma_{jk}(x)- - \sigma_{jk}(x')),$$

$$c_3(x,x',p,\lambda) = \int_{\mathbb{R}_*^m} [(\lambda+|x-x'+\gamma(x,\zeta)-\gamma(x',\zeta)|^2)^{p/2} - - p(\lambda+|x-x'|^2)^{p/2-1} \sum_{i=1}^{d} (x_i-x_i')(\gamma_i(x,\zeta)- - \gamma(x',\zeta))]\pi(d\zeta),$$

we obtain for every $t \geq 0$, x in $\mathbb{R}^d$, $\nu(.)$ additive control

$$E \{ (\lambda+|y_x(t)-y_{x'}(t)|^2)^{p/2}\exp(-\beta'(p,\lambda)t) \leq (\lambda+|x-x'|^2)^{p/2}, \quad (1.14)$$

and if

$$\beta_p = \limsup_{\lambda \to 0} \beta'(p,\lambda) \quad (1.15)$$

then

$$E \{|y_x(t)-y_{x'}(t)|^p\exp(-\beta_p t) \leq |x-x'|^p, \quad \forall t\geq 0, \quad \forall x,x'\in\mathbb{R}^d, \quad (1.16)$$

for any $p > 0$. Remark that under the assumption (1.1), $\beta_p(\lambda)$ is finite for every $\lambda > 0$ and if

$$\left.\begin{aligned} &|g(x)-g(x')|^p+|\sigma(x)-\sigma(x')|^p+ \int_{\mathbb{R}_*^m}|\gamma(x,\zeta)-\gamma(x',\zeta)|^p\pi(d\zeta)\leq \\ &\leq C|x-x'|^{p-\varepsilon} \quad , \end{aligned}\right\} \quad (1.17)$$

for every $p \geq 2$, x,x' in $\mathbb{R}^d$ and some positive constants $C=C(p)$, $\varepsilon=\varepsilon(p)$ then we have for every fixed $p > 0$,

$$\beta'(p,\lambda) \to 0 \quad \text{as} \quad \lambda \to \infty . \tag{1.18}$$

On the other hand, by means of the hypothesis (1.2) we can show that β_p, given by (1.15), is finite.

2. A NONLINEAR SEMIGROUP

We say that $(\nu(t),\ t \geq 0)$ is an admissible control if there exists a version of the process $(\nu(t,\omega),\ t \geq 0,\ \omega \in \Omega)$ such that

$$\left.\begin{array}{l} \nu(t,\omega)=\nu^+(t,\omega)-\nu^-(t,\omega),\ \forall t \geq 0,\ \forall\omega \in \Omega, \\ \nu^+(t,\omega) \text{ and } \nu^-(t,\omega) \text{ are progressively measurable} \\ \text{processes with finite moments, right continuous having} \\ \text{left-hand limits, non-negative and non-decreasing by} \\ \text{coordinates.} \end{array}\right\} \tag{2.1}$$

In general, we have $\nu^+ = (\nu_i^+,\ i=1,\ldots d)$,

$$\nu_i^+(t) = \nu_i^+(0)+\sup \{ \sum_{j=1}^{n} (\nu_i(t_j)-\nu_i(t_{j-1}))^+ : 0=t_0<t_1<\ldots<t_n=t\}, \tag{2.2}$$

where $(.)^+$ in the right-hand side denotes the positive part of a real number, and a similar definition for the process ν^-, replacing the positive part by the negative part. Under this notation, the total variation process is given by

$$\eta(t,\omega) = \nu^+(t,\omega) + \nu^-(t,\omega) \ ,\ \forall t \geq 0,\ \forall\omega \in \Omega . \tag{2.3}$$

A control $(\nu(t), t \geq 0)$ is said to be a Lipschitz control if for some bounded and progressively measurable process $(\dot{\nu}(t),\ t \geq 0)$ we have

$$\nu(t,\omega)=\nu(0,\omega) + \int_0^t \dot{\nu}(s,\omega)ds,\ \forall t \geq 0,\ \forall\omega \in \Omega . \tag{2.4}$$

On the other hand, a control $(\nu(t),\ t \geq 0)$ is called an impulse control if it exists two sequences $(\theta_j,\ j=0,1,\ldots)$, and $(\xi_j,\ j=0,1,\ldots)$ of stopping times and random variables respectively, such that

$$\left.\begin{aligned}
&\nu(t,\omega)=\sum_{j=0}^{\infty} \xi_j(\omega)\chi(\theta_j(\omega)\le t),\ \forall t \ge 0,\ \forall\omega \in \Omega ,\\
&0 = \theta_0 < \theta_j < \theta_{j+1} ,\ \text{if } \theta_j < +\infty ,\\
&\theta_j(\omega) \to +\infty \text{ as } j \to \infty ,\ \forall\omega \in \Omega\\
&\xi_j \text{ is } F_{\theta_j} \text{ measurable} ,\ \forall j = 0,1,\ldots,
\end{aligned}\right\} \quad (2.5)$$

with χ being the characteristic function. We denote by V, V_ℓ and V_i the set of all controls, Lipschitz controls and impulse controls, respectively.

THEOREM 2.1

Let the assumptions (1.1), (1.2) and (1.8) hold. Then the infimum of the payoff functional (1.7) is the same over the sets V, V_ℓ, V_i of admissibles controls, Lipschitz control and impulse control, respectively.

Outline of the proof.

To each admissible control $(\nu(t),\ t \ge 0)$ we can associate a Lipschitz control $(\nu^\ell(t),\ t \ge 0)$,

$$\left.\begin{aligned}
&\nu^\ell(t) = \nu_\ell^+(t) - \nu_\ell^-(t),\\
&\nu_\ell^\pm(t) = \begin{cases} (\varepsilon-t)\nu^\pm(0)\varepsilon^{-1} + \varepsilon^{-1}\displaystyle\int_0^t (\delta\wedge\nu^\pm)(s)\,ds & \text{if } 0\le t\le\varepsilon,\\ \varepsilon^{-1}\displaystyle\int_{t-\varepsilon}^t (\delta\wedge\nu^\pm(s))ds, & \text{otherwise,}\end{cases}
\end{aligned}\right\} \quad (2.6)$$

where $\varepsilon,\delta > 0$ and $\delta \wedge \nu^\pm(s)$ is taken by coordinates.

It is clear that ν^ℓ belongs to V_ℓ and $\nu^\ell(t)$ converges to $\nu(t-)$ as ε,δ tend to 0, ∞. Next, given a ν^ℓ in V_ℓ, we define an impulse control $(\nu^i(t),\ t \ge 0)$ by

$$\left.\begin{aligned}
&\nu^i(t) = \nu^\ell(t_j) ,\ \text{if } t_j\le t<t_{j+1} ,\\
&0=t_1<t_2<\ldots<t_j,\ t_j \to \infty,\ t_{j+1}-t_j \le \varepsilon .
\end{aligned}\right\} \quad (2.7)$$

Then, as ε goes to zero the process ν^i approach ν^ℓ. Finally, we conclude by proving that the state $(y(t,\nu),\ t \ge 0)$ varies continuously with respect to the control. □

In order to define the so-called nonlinear semigroup associated with the model (1.4),...,(1.7), we shall include in our control the whole Wiener-Poisson space $(\Omega, F, P, F_t, w_t, \mu_t, t \geq 0)$ and the processes $(y(t), t \geq 0)$, $(\nu(t), t \geq 0)$, $(\eta(t), t \geq 0)$ related by (1.4), (1.6). These sets are refered to as admissible systems Λ. Note that the Levy measure π and the coefficients g, σ, γ are fixed.

Similarly, we call Λ an r-admissible system if the corresponding process $(\nu(t), t \geq 0)$ is a Lipschitz control satisfying $\dot{\nu}=(\dot{\nu}_i, i=1,\ldots,d)$

$$-r \leq \dot{\nu}_I(t,\omega) \leq r \ , \ \forall t \geq 0, \ \forall \omega \in \Omega \ , \tag{2.8}$$

where r is a positive parameter sufficiently large. For such a system, the stochastic differential equation (1.4) and the equality (1.6) become

$$\left.\begin{aligned} dy(t) &= (g(y(t)) + \dot{\nu}(t))dt + \sigma(y(t))dw_t + \\ &\qquad \int_{\mathbb{R}^m_*} \gamma(y(t),\zeta)d\mu_t(\zeta), \ t \geq 0 \\ d\eta(t) &= |\dot{\nu}(t)|dt \ , \ t \geq 0 \\ y(0) &= x \quad , \quad \eta(0) = \xi \ , \end{aligned}\right\} \tag{2.9}$$

where $|\dot{\nu}(t)|$ denotes the absolute value taken by coordinates.

Let $S_p(\mathbb{R}^d \times \mathbb{R}^d_+)$ be the convex cone of $B_p(\mathbb{R}^d \times \mathbb{R}^d_+)$ composed by all positive and upper semicontinuous functions. The set $B_p(\bar{O})$ denotes the Banach space of all Borel measurable functions $h(z)$ from $\bar{O}$, a closed subset of $\mathbb{R}^n$, into $\mathbb{R}$ satisfying

$$|h(z)| \leq C(1+|z|^p) \quad , \quad \forall z \ , \tag{2.10}$$

for some constant $C > 0$. For a fixed $\lambda > 0$, the norm is given by

$$\| h\|_p = \sup\{ \ |h(z)|(1+|z|^2)^{-p/2} \ : \ z \ \} \ . \tag{2.11}$$

Define

$$\left.\begin{aligned} J_{x\xi}(\Lambda,h,t) = E\{&\int_0^t [f(y(s))+\alpha c(\eta(s))]e^{-\alpha s}\, ds + \\ &\qquad + h(y(t-),\eta(t))e^{-\alpha t} \},t \geq 0, \end{aligned}\right\} \tag{2.12}$$

$$Q_r(t)h(x,\xi)=\inf\{ J_{x\xi}(\Lambda,h,t) \ : \ \Lambda \text{ r-admissible system} \}, \tag{2.13}$$

and

$$Q(t)h = \lim_{r\to\infty} Q_r(t)h \quad , \quad t \geq 0 \ , \tag{2.14}$$

being a decreasing limit. Note that in view of Theorem 2.1, if h is lower semicontinuous then

$$Q(t)h(x,\xi)=\inf\{ J_{x\xi}(\Lambda,h,t) \ : \ \Lambda \text{ admissible system} \}. \tag{2.15}$$

THEOREM 2.2

Under the hypotheses of Theorem 2.1 and assuming that

$$f,\ c \text{ are continuous,} \tag{2.16}$$

the family of operators $(Q(t),\ t \geq 0)$ *is a nonlinear semigroup on* $S_p(\mathbb{R}^d \times \mathbb{R}^d_+)$ *enjoying the following properties :*

$$\|Q(t)h\|_p \leq \alpha_0^{-1} \| f\|_p + \|c\|_p + \|h\|_p \ , \tag{2.17}$$

$$\text{if } h_1, h_2 \in S_p(\mathbb{R}^d \times \mathbb{R}^d_+),\ h_1 \leq h_2 \text{ then } Q(t)h_1 \leq Q(t)h_2,\ \forall t \geq 0, \tag{2.18}$$

$$\text{if } (h_n, n=0,1,\ldots) \subset S_p(\mathbb{R}^d \times \mathbb{R}^d_+),\ h_n \downarrow h_0 \text{ then } Q(t)h_n \downarrow Q(t)h_0, \tag{2.19}$$

$$\left.\begin{array}{l} \text{if } h \in S_p(\mathbb{R}^d \times \mathbb{R}^d_+) \text{ then } Q(t)h \in S_p(\mathbb{R}^d \times \mathbb{R}^d_+) \text{ and} \\ Q(s)Q(t)h = Q(s+t)h, \quad \forall s,t \geq 0. \end{array}\right\} \tag{2.20}$$

Moreover, the value function

$$\hat{u}(x,\xi) = \inf \{ J_{x\xi}(\Lambda,0,\infty) \ : \ \Lambda \text{ admissible system} \}, \tag{2.21}$$

is the maximum solution of the problem :

$$\text{find } u \text{ in } S_p(\mathbb{R}^d \times \mathbb{R}^d_+) \text{ such that } Q(t)u = u \ , \ \forall t \geq 0 \ . \tag{2.22}$$

Furthermore, the function $\hat{u}(x,\xi)$ *coincides with the optimal cost (1.7) when* $\xi = 0$.

Outline of the proof

By using classical techniques (e.g. Nisio [26] , Bensoussan and Lions [4] , Lions and Menaldi [21]), we establish that the semigroup property (2.20) is verified for the family $(Q_r(t),\ t \geq 0)$ when f, h are bounded. Next, by means of the estimate

$$E\{\ (\lambda + |y_x(t)|^2)^{q/2} e^{-\alpha t}\ \} \ \leq\ (\lambda + |x|^2)^{q/2} e^{-\alpha_0 t}, \tag{2.23}$$

valid for some $\lambda = \lambda(q,r)$ and any x, t, q, we obtain (2.17),...,(2.20) for $Q_r(t)$ in lieu of $Q(t)$.

Finally, we show that the function

$$\hat{u}_r(x,\xi) = \inf\{\ J_{x\xi}(\Lambda,0,\infty)\ :\ \Lambda \text{ r-admissible system }\} \tag{2.24}$$

is the unique stationary point of $(Q_r(t),\ t \geq 0)$ and then we conclude.

□

REMARK 2.1

Note that (2.9), (2.12), (2.24) define a classical stochastic control problem. The corresponding Hamilton-Jacobi-Bellman equation to be satisfied by $\hat{u}_r(x,\xi)$ is

$$Au + r \sum_{i=1}^{d} B_i(u) = h \quad \text{in } \mathbb{R}^d \times \mathbb{R}^d_+ , \tag{2.25}$$

where

$$\left.\begin{aligned} &h(x,\xi) = f(x) + \alpha\, c(\xi), \\ &B_i(u) = -\min\{\theta \frac{\partial u}{\partial x_i} + |\theta| \frac{\partial u}{\partial \xi_i}\ :\ -1 \leq \theta \leq 1\}, \end{aligned}\right\} \tag{2.26}$$

and A is the following integro-differential operator

$$\left.\begin{aligned} Av(x) = &-\frac{1}{2} \sum_{i,j=1}^{d} \Big(\sum_{k=1}^{n} \sigma_{ik}(x)\sigma_{jk}(x)\Big) \frac{\partial^2 v}{\partial x_i \partial x_j}(x) - \sum_{i=1}^{d} g_i(x) \frac{\partial v}{\partial x_i}(x) + \\ &+ \alpha\, v(x) - \int_{\mathbb{R}^m_*} [v(x + \gamma(x,\zeta)) - v(x) - \\ &- \sum_{i=1}^{d} \gamma_i(x,\zeta) \frac{\partial v}{\partial x_i}(x)\ ]\ \pi(d\zeta). \end{aligned}\right\} \tag{2.27}$$

Therefore, as r goes to infinity we obtain

$$\left.\begin{aligned} & Au \le h \quad \text{in } \mathbb{R}^d \times \mathbb{R}^d_+ , \\ & \left|\frac{\partial u}{\partial x_i}\right| \le \frac{\partial u}{\partial \xi_i} \quad \text{in } \mathbb{R}^d \times \mathbb{R}^d_+ , \quad \forall i = 1,\ldots,d, \end{aligned}\right\} \tag{2.28}$$

i.e., the Hamilton-Jacobi-Bellman conditions associated with the value function (2.21). □

Let $K_p(\bar{O})$ be the convex cone of $B_p(\bar{O})$ composed by all positive functions h(z) satisfying

$$h(z) \ge c|z|^p - C \quad , \quad \forall x \in \bar{O} \subset \mathbb{R}^n, \tag{2.29}$$

for some constants $C \ge c > 0$. The Banach space $C_p^o(\bar{O})$ is the set of all functions h(z) such that :

$$\left.\begin{aligned} & \text{for every } \varepsilon > 0 \text{ there exists a constant } C = C(\varepsilon) \\ & \text{verifying for all } z,\ z' \text{ in } \bar{O} , \\ & |h(z) - h(z')| \le \varepsilon(1+|z|^p) + C|z-z'|^p , \end{aligned}\right\} \tag{2.30}$$

with the norm $\|.\|_p$ of $B_p(\bar{O})$. In what follows, we take either $\bar{O} = \mathbb{R}^d$ or $\bar{O} = \mathbb{R}^d \times \mathbb{R}^d_+$, and $\bar{O}$ denotes the closure or the set O .

Suppose that for $p \ge 0$,

$$\left.\begin{aligned} & f \in C_p^o(\mathbb{R}^d), \ c \in C_p^o(\mathbb{R}^d_+) , \\ & f,\ c \ge 0 \quad , \ c \in K_p(\mathbb{R}^d_+) , \end{aligned}\right\} \tag{2.31}$$

and for the same p, $\alpha > 0$ if $p = 0$, and if $p > 0$,

$$\alpha \ge \alpha_o + \max\{\beta_q, \beta(p,\lambda), \beta'(q+(p-q)r,\ \lambda)\} \tag{2.32}$$

for some constants $0<q\le p$, $r > 1$, $\lambda > 0$, $\alpha_o > 0$ and the notation (1.9), (1.13) and (1.15). Note that if (1.17) holds then (2.32) is satisfied for every $\alpha > 0$.

THEOREM 2.3

Assume the conditions (1.1), (1.2), (2.31) and (2.32) hold. Then the value function $\hat{u}(x,\xi)$ given by (2.21) belongs to $C_p^o(\mathbb{R}^d \times \mathbb{R}_+^d)$ and the dynamic programming equation is valid, i.e. for every x,ξ,

$$\hat{u}(x,\xi) = \inf \{J_{x\xi}(\Lambda,\hat{u},\theta) : \Lambda \text{ admissible system }\}, \tag{2.33}$$

where θ is any stopping time associated with Λ.

Outline of the proof

Based on the last condition of (2.31) we show that there exists a constant $C > 0$ independent of x,ξ,ν such that the infimum (2.21) can be restricted to those controls ν satisfying

$$E\{ \int_0^\infty |\eta(t)|^p e^{-\alpha t} dt \} \le C(1+|x|^p + |\xi|^p), \tag{2.34}$$

where η is the variation process (2.3). This implies

$$E\{ \int_0^\infty |y(t)|^p e^{-\alpha t} dt \} \le C(1+|x|^p+|\xi|^p), \tag{2.35}$$

for another constant $C > 0$.

Next, by means of estimates (1.14), (1.16), which are valid for the controlled processes $y_x(t)$, $y_{x'}(t)$, we can prove that the value function $\hat{u}(x,\xi)$ belongs to $C_p^o(\mathbb{R}^d \times \mathbb{R}_+^d)$ after using Hölder's inequality and (2.34), (2.35).

The dynamic programming equation (2.33) follows from Theorem 2.2 and (2.15). □

REMARK 2.2

We may replace the assumption (2.31) by

$$f \in C_p^o(\mathbb{R}^d) \cap K_p(\mathbb{R}^d), \quad c \in C_p^o(\mathbb{R}_+^d) \cap K_p(\mathbb{R}_+^d), \tag{2.36}$$

for some given p, $q \ge 0$. In this case, if (2.32) holds for $p \vee q$ instead of p, then the value function (2.21) belongs to $C_{p,q}^o(\mathbb{R}^d \times \mathbb{R}_+^d)$, Theorem 2.2 is valid with obvious modifications and (2.33) remains true. To define the space $C_{p,q}^o(\mathbb{R}^d \times \mathbb{R}_+^d)$ we replace (2.30) by :

$$\left.\begin{array}{l} \text{for every } \varepsilon > 0 \text{ there exists a constant } C=C(\varepsilon) \text{ verifying} \\ \text{for all } x,\ x' \text{ in } \mathbb{R}^d,\ \xi,\ \xi' \text{ in } \mathbb{R}^d_+, \\ |h(x,\xi)-h(x',\xi')| \le \varepsilon(1+|x|^p+|\xi|^q)+C(|x-x'|^p+|\xi-\xi'|^q), \end{array}\right\} \quad (2.37)$$

with a weighted norm $\|.\|_{p,q}$. □

3. VARIATIONAL FORMULATION

Let $W^{1,\infty}_{p,q}(\mathbb{R}^d\times\mathbb{R}^d_+)$, $p,q \ge 0$, be the space of all locally Lipschitz continuous functions v from $\mathbb{R}^d \times \mathbb{R}^d_+$ into $\mathbb{R}$ satisfying

$$\left.\begin{array}{l} |v(x,\xi)-v(x',\xi')| \le C(|x-x'|^p+|\xi-\xi'|^q) + \\ \qquad + C(1+|x|^p+|\xi|^q)^{1/p'}|x-x'|+ \\ \qquad + C(1+|x|^p+|\xi|^q)^{1/q'}|\xi-\xi'|, \end{array}\right\} \quad (3.1)$$

for some constant $C > 0$, every $x,x',\xi,\ \xi'$ and $p'=\infty$ if $0\le p\le 1$, $p' = p/(p-1)$ otherwise and a similarly definition for q'.

Under suitable conditions and if A denotes the integro-differential operator (2.27), we can look at Av as a distribution on $\mathbb{R}^d$. For instance, assuming (1.1), (1.2) and

$$\int_{\mathbb{R}^m_*} |\gamma(x,\zeta)|\pi(d\zeta) \le C_1(1+|x|) \quad (3.2)$$

for every x in $\mathbb{R}^d$ and some constant C_1, we have for every infinitely differentiable function φ with compact support on $\mathbb{R}^d$ and any v in $W^{1,\infty}_p(\mathbb{R}^d)$,

$$\left.\begin{array}{l} \langle Av, \varphi\rangle = \displaystyle\sum_{i,j=1}^{d} \int_{\mathbb{R}^d} \left(\frac{\partial v}{\partial x_i}(x)\right)\left[\frac{\partial}{\partial x_j}(a_{ij}(x)\varphi(x)\right]dx + \\ \qquad + \displaystyle\sum_{i=1}^{d} \int_{\mathbb{R}^d} \tilde{a}_i(x)\left(\frac{\partial v}{\partial x_i}(x)\right)\varphi(x)dx + \int_{\mathbb{R}^d} \alpha v(x)\varphi(x)dx + \\ \qquad - \displaystyle\int_{\mathbb{R}^d} \varphi(x)dx \int_{\mathbb{R}^m_*} [v(x+\gamma(x,\zeta))-v(x)\ ]\pi(d\zeta)\ , \end{array}\right\} \quad (3.3)$$

with

$$\left.\begin{aligned} a_{ij}(x) &= \frac{1}{2}\sum_{k=1}^{n} \sigma_{ik}(x)\sigma_{jk}(x), \\ \tilde{a}_i(x) &= -g_i(x) + \int_{\mathbb{R}^m_*} \gamma_i(x,\zeta)\pi(d\zeta),\ i,j=1,\ldots d. \end{aligned}\right\} \qquad (3.4)$$

Note that (3.2) means that the nonlocal part of A, i.e. the integral part, is a first order operator. By the way, the space $W^{1,\infty}_p(\mathbb{R}^d)$ is defined similarly to $W^{1,\infty}_{p,q}(\mathbb{R}^d \times \mathbb{R}^d_+)$ without the variable ξ.

Consider the problem :

$$\left.\begin{aligned} &\text{find } u(x,\xi) \text{ in } W^{1,\infty}_{p,q}(\mathbb{R}^d \times \mathbb{R}^d_+) \text{ such that} \\ &Au(.,\xi) \le f + \alpha c(\xi) \text{ in } \mathcal{D}'(\mathbb{R}^d),\ \forall\xi \in \mathbb{R}^d_+, \\ &\left|\frac{\partial u}{\partial x_i}\right| \le \frac{\partial u}{\partial \xi_i} \quad \text{a.e. in } \mathbb{R}^d \times \mathbb{R}^d_+ ,\quad \forall i=1,\ldots,d, \end{aligned}\right\} \qquad (3.5)$$

where $\mathcal{D}'(\mathbb{R}^d)$ denotes the space of Schwartz' distributions.

Suppose that for $p,q \ge 0$

$$f \in W^{1,\infty}_p(\mathbb{R}^d) \cap K_p(\mathbb{R}^d) \ ,\ c \in W^{1,\infty}_q(\mathbb{R}^d_+) \cap K_q(\mathbb{R}^d_+) \qquad (3.6)$$

and for $s = p \vee q$,

$$\left.\begin{aligned} &\alpha \ge \alpha_0 + \max\{\beta_1,\ \beta(s,\lambda)\} \quad ,\ \text{if } 0\le s\le 1 , \\ &\alpha \ge \alpha_0 + \max\{\beta_r,\ \beta(s+t,\lambda)\ ,\ \beta'(s+t,\lambda)\} \ ,\ \text{if } s > 1, \\ &\text{for some constants } \lambda > 0,\ \alpha_0 > 0,\ 1<r<s, t = (s-r/(r-1). \end{aligned}\right\} \qquad (3.7)$$

If $q \ge p$ then f needs not to belong to $K_p(\mathbb{R}^d)$, i.e. to satisfy (2.29). When (1.10) and (1.17) hold, we see that (3.7) reduces to

$$\alpha > \beta_1 \ ,\ \text{given by (1.15) with } p = 1. \qquad (3.8)$$

THEOREM 3.1

Under the assumptions (1.1), (1.2), (3.2), (3.6) and (3.7) the problem (3.5) admits a maximum solution $\hat{u}(x,\xi)$, which is given explicitly as the value function (2.21).

Outline of the proof

Using the penalized problem (2.25) we can show that the value functions $\hat{u}_r(x,\xi)$, given by (2.24), is the maximum solution of the problem :

$$\begin{aligned} &\text{find } u_r \text{ in } W^{1,\infty}_{p,q}(\mathbb{R}^d \times \mathbb{R}^d_+) \text{ such that} \\ &Au_r + r \sum_{i=1}^{d} B_i(u_r) = h \text{ in } \mathcal{D}'(\mathbb{R}^d \times \mathbb{R}^d_+), \end{aligned} \tag{3.9}$$

where

$$\begin{aligned} &h(x,\xi) = f(x) + \alpha\, c(\xi) , \\ &B_i(u) = -\min \{\theta \frac{\partial u}{\partial x_i} + |\theta| \frac{\partial u}{\partial \xi_i} : -1 \le \theta \le 1 \} . \end{aligned} \tag{3.10}$$

For classical arguments we refer to Fleming-Rishel [11] and Krylov [20].

At the same time, we obtain a priori estimate for the functions $\hat{u}_r, r\ge 1$, i.e. the condition (3.1) holds for $\hat{u}_r$ uniformly in $r \ge 1$.

Since $\hat{u}_r$ converges to $\hat{u}$ as r tends to infinity, we conclude by passing to the limit and noting that $\hat{u}_r \ge \hat{u}$ for every r. □

REMARK 3.1

A variational formulation similar to (3.5) can be considered in the space $C^{o}_{p,q}(\mathbb{R}^d \times \mathbb{R}^d_+)$. In this case the operator A is used in the martingale sense, the last inequality of (3.5) becomes

$$u(x,\xi) \le u(x\pm\xi',\xi+\xi') , \quad \forall x \in \mathbb{R}^d, \ \forall \xi,\xi' \in \mathbb{R}^d_+ , \tag{3.11}$$

and the assumption (3.2) is removed. □

In order to give a charaterization of the value function as a unique solution of variational inequality instead of a maximum subsolution, cfr. (3.5), we rewrite the integro-differential operator (2.27) as follows

$$
\left.\begin{aligned}
Av(x) = &- \sum_{i,j=1}^{d} a_{ij}(x)\frac{\partial^2 v}{\partial x_i \partial x_j}(x) + \sum_{i=1}^{d} a_i(x)\,\frac{\partial v}{\partial x_i}(x) + \\
&+ \alpha\, v(x) - \int_{\mathbb{R}^d_*}[v(x+z)-v(x) - \\
&- \sum_{i=1}^{d} z_i \frac{\partial v}{\partial x_i}(x)]\; b(x,z)m(dz),
\end{aligned}\right\} \quad (3.12)
$$

where $m(dz)$ denotes a Ra don measure on $\mathbb{R}^d_*$ and $b(x,z)$ is a Borel measurable function from $\mathbb{R}^d \times \mathbb{R}^d_*$ into $[0,\infty]$ such that for some positive constants b_0, b_1 and any $p \geq 2$, z in $\mathbb{R}^d_*$ we have

$$
\left.\begin{aligned}
&0 \leq b(x,z) \leq b_0\;(1+|x|^2), \\
&\left|\frac{\partial b}{\partial x_i}(x,z)\right| \leq b_1\;(1+|x|)\;, \quad \forall i = 1,\ldots,d, \\
&\int_{\mathbb{R}^d_*} |z|^p m\,(dz) < \infty
\end{aligned}\right\} \quad (3.13)
$$

for every x in $\mathbb{R}^d$. Note that $b(x,z)$, $m(dz)$ and $\gamma(s,\zeta)$, $\pi(d\zeta)$ are related each other by

$$
\pi(\{\zeta\; :\; \gamma(x,\zeta) \in B\}) = \int_B b(x,z)m(dz), \qquad (3.14)
$$

for any Borel measurable subset B of $\mathbb{R}^d_*$. Also

$$
\left.\begin{aligned}
&a_{ij}(x) = \frac{1}{2}\sum_{k=1}^{n} \sigma_{ik}(x)\;\sigma_{jk}(x)\;, \\
&a_i(x) = -g_i(x)\quad, \quad \forall x \in \mathbb{R}^d\;, \;\forall i,j=1,\ldots d.
\end{aligned}\right\} \quad (3.15)
$$

Assuming (1.1), (1.2), (3.13), (3.14), (3.15) and

$$
\left.\begin{aligned}
&\left|\frac{\partial^2 b}{\partial x_\ell \partial x_h}(x,z)\right| \leq b_2 \quad , \quad \left|\frac{\partial^2 a_{ij}}{\partial x_\ell \partial x_h}(x)\right| \leq C, \\
&\forall i,j,\ell,h=1,\ldots d,\ \text{a.e. } x \text{ in } \mathbb{R}^d,\ \forall z \in \mathbb{R}^d,
\end{aligned}\right\} \quad (3.16)
$$

for some positive constants b_2, C, then we can consider Av as a distribution on $\mathbb{R}^d$ for any v belonging to $B_p(\mathbb{R}^d)$.

Let H, V be the following weighted Sobolev spaces :

$$\left.\begin{aligned} & v \in H \text{ iff } vq_0 \in L^2(\mathbb{R}^d) , \\ & v \in V \text{ iff } v \in H,\ (\frac{\partial v}{\partial x_i})q_1 \in L^2(\mathbb{R}^d),\ i=1,\ldots,d, \\ & q_0(x) = (1+|x|^2)^{-k},\ q_1(x)=q_0(x)(1+|x|^2)^{1/2} ,\ k=d+1+p/2, \end{aligned}\right\} \quad (3.17)$$

with the natural Hilbertian norms

$$\left.\begin{aligned} & |v| = \|v\|_H = (\int_{\mathbb{R}^d} |v(x)q_0(x)|^2 dx)^{1/2} , \\ & \|v\| = \|v\|_V = (\|v\|_H^2 + \sum_{i=1}^{d} \int_{\mathbb{R}^d} |(\frac{\partial v}{\partial x_i}(v))q_1(x)|^2 dx)^{1/2}. \end{aligned}\right\} \quad (3.18)$$

We have the classical continuous inclusions with dense image

$$V \subset H = H' \subset V' \qquad (3.19)$$

for the dual spaces H', V'. Under the assumptions (1.1), (1.2), (3.13), (3.14) and (3.15), the integro-differential operator A may be regarded as a linear and bounded operator from V into V'. Moreover if (3.16) holds then we can find a constant β such that

$$\langle Av, v\rangle \geq (\alpha-\beta) |v|^2 , \quad \forall v \in V. \qquad (3.20)$$

Thus we suppose α sufficiently large, i.e.

$$\alpha - \beta = \alpha_0 > 0. \qquad (3.21)$$

This means that A defines a continuous bilinear form on V × V, which is coercive only under the norm H.

Assume the function $c(\xi)$ satisfies

$$c(\xi) = c_0 + \sum_{i=1}^{d} c_i\xi_i , \quad \forall\xi = (\xi_1,\ldots,\xi_d) \in \mathbb{R}_+^d , \qquad (3.22)$$

and define the convex closed subset K of V by

$$K = \{v \in V : |\frac{\partial v}{\partial x_i}| \leq c_i \ , \text{ a.e. in } \mathbb{R}^d, \ \forall i=1,\ldots,d \}. \quad (3.23)$$

Since $c(\xi)$ is non-negative, it results $c_i \geq 0$, for any i. Therefore, if $c_i = 0$ then $v \in K$ implies that v is independent of the variable x_i. Consider the variational inequality :

$$\left. \begin{array}{l} \text{find } u \text{ in } K \text{ such that} \\ \langle Au, v-u\rangle \geq (f,v-u) \ , \ \forall v \in K, \end{array} \right\} \quad (3.24)$$

where $(.,.)$ denotes the inner product in V.

THEOEREM 3.2
Let the assumptions (1.1), (1.2), (3.13), (3.16), (3.22) and

$$f, c \geq 0, \ f \in W_p^{1,\infty}(\mathbb{R}^d), \quad \alpha \text{ sufficiently large} \quad (3.25)$$

then the variational inequality (3.24) possesses a unique solution u(x), which belongs to $W_p^{1,\infty}(\mathbb{R}^d)$ *and is given explicilty by*

$$u(x) = \hat{u}(x) - c_0 \quad , \quad \forall x \text{ in } \mathbb{R}^d, \quad (3.26)$$

where $\hat{u}(x)$ *is the optimal cost (1.7).*

Outline of the proof
As in Theorem 3.1, we use the penalized problem (3.9). In this case, the equation

$$Au^{\varepsilon} + \frac{1}{\varepsilon} \sum_{i=1}^{d} (|\frac{\partial u^{\varepsilon}}{\partial x_i}| - c_i)^+ = f \text{ in } V', \quad (3.27)$$

has one and only one solution, which is given explicitly by

$$u^{\varepsilon}(x) = \hat{u}_r(x,0) - c_0 \ , \ r = \frac{1}{\varepsilon} \quad , \quad (3.28)$$

for any $\varepsilon > 0$, x in $\mathbb{R}^d$.

Since the family $(u^{\varepsilon},\ 0<\varepsilon\le 1)$ stays in a bounded subset of $W_p^{1,\infty}(\mathbb{R}^d)$, we can take the limit in (3.27). The uniqueness follows from (3.20).□

REMARK 3.2

Noting that the convex set K is bounded in V, we can extend the existence and uniqueness of solution for the variational inequality (2.24) for any f in V'. However, to have the representation (3.26) we need f in $C_p^o(\mathbb{R}^d)$.□

REMARK 3.3

A problem similar to (3.24) can be considered for more general functions $c(\xi)$. That formulation use weighted Sobolev spaces on $\mathbb{R}^d \times \mathbb{R}_+^d$. □

It is clear now that by means of the results of sections 2 and 3, we can establish the connection among this problems and the so-called cheap control problems, monotone follower problems and switching type problems. In particular, the limits in (8) and (9) can be found rigorously. That is presisely the origin of this class of singular control problems.

On the other hand, we can add a condition :

$$\eta = \nu \text{ is increasing and non-negative,}$$

before taking the infimum in (2.21). Most of the results remain valid.

4. ONE DIMENSIONAL CASE

The additive control problem in $\mathbb{R}^d$ can be reduced to an one-dimensional problem if, for instance

$$\left.\begin{aligned} & g_i(x) = g_i(x_i) \quad , \quad \gamma_i(x,\zeta) = \gamma_i(x_i,\zeta) \ , \ \forall i \ , \\ & \sigma_{ik} = \begin{cases} \sigma_{ii}(x_i) & , \text{ if } i=k \\ 0 & , \text{ otherwise,} \end{cases} \quad , \ \forall i,\ k, \\ & f(x) = \sum_{i=1}^{d} f_i(x_i) \quad , \quad c(\xi) = \sum_{i=1}^{d} c_i(\xi_i). \end{aligned}\right\} \tag{4.1}$$

Thus

$$\hat{u}(x,\xi) = \sum_{i=1}^{d} \hat{u}_i(x_i,\ \xi_i),$$

where $\hat{u}_i(x_i,\xi_i)$ is the value function associated with an one-dimensional problem, where the data are $g_i(x_i)$, $\sigma_{ii}(x_i)$, $\gamma_i(x_i,\zeta)$, $f_i(x_i)$, $c_i(\xi_i)$.

Another case is the isotropic process, i.e. when $f(x)$, $c(\xi)$ have spherical symmetry and some other conditions are required for the coefficients g,σ,γ.

Notice that the ε-switching cost impulse control problem (9) is not decomposable with the assumptions (4.1), while the limit problem can be decomposed.

More results are available for one-dimensional models, in particular with respect to the existence of an optimal control. This is due to the fact that, under suitable assumptions, the boundary of the continuation set is reduced to one or two points, and therefore no regularity question is involved for that boundary. We point out here that most of the results that can be found in the literature deal with one-dimensional process and most of the time with Brownian motion.

To simplify this presentation, we assume

$$\left.\begin{array}{l} g(x),\ \sigma(x) \text{ and } \gamma(x,\xi) \text{ Lipschitz continuous (in } x) \\ \text{and bounded functions from } \mathbb{R} \text{ and } \mathbb{R}\times\mathbb{R}_* \text{ into } \mathbb{R}, \text{ respectively} \end{array}\right\} \quad (4.3)$$

$$\int_{\mathbb{R}_*} \pi(d\zeta) < \infty \quad , \quad \sigma(x) \ge \sigma_0 > 0,\ \forall x \in \mathbb{R}\ , \quad (4.4)$$

$$c|x|^2 - C \le f(x) \le C(1+|x|^2) \quad , \quad \forall x \in \mathbb{R} \quad (4.5)$$

for some constants $C \ge c > 0$. Only the monotone follower case will be considered, i.e. the optimal cost is given by

$$\left.\begin{array}{l} \hat{u}(x) = \inf\{\ J_x(\nu) : \nu \text{ increasing and non-negative } \}, \\ J_x(\nu) = E\{\ \int_0^\infty f(y_x(t))e^{-\alpha t}\, dt\ \}\ , \end{array}\right\} \quad (4.6)$$

and

$$\left.\begin{array}{l} y(t) = x + \int_0^t g(y(s))ds + \int_0^t \gamma(y(s))dw_s\ + \\ \qquad + \int_0^t \int_{\mathbb{R}_*} \gamma(y(s),\zeta)d\mu_s(\zeta) + \nu(t). \end{array}\right\} \quad (4.7)$$

Then, with

$$\left.\begin{aligned} Av(x) &= -\frac{1}{2}\sigma^2(x)\frac{d^2v}{dx^2}(x) - g(x)\frac{dv}{dx}(x) + \\ &+ \alpha\, v(x) - \int_{\mathbb{R}_*}[v(x+\gamma(x,\zeta))-v(x) - \\ &- \gamma(x,\zeta)\frac{dv}{dx}(x)]\pi(d\zeta). \end{aligned}\right\} \quad (4.8)$$

The variational inequality for the optimal cost function $\hat{u}$ is, at least if u is smooth

$$\left.\begin{aligned} &A\hat{u} \le f \quad , \quad \hat{u}' \ge 0, \\ &(A\hat{u} - f)\hat{u}' = 0 \quad , \end{aligned}\right\} \quad (4.9)$$

where $\hat{u}'$ denotes the first derivative of $\hat{u}$.

THEOREM 4.1

Under the assumptions (4.3), (4.4), (4.5) and

$$f \in W_2^{1,\infty}(\mathbb{R}), \text{ cfr. } (3.1), \quad (4.10)$$

there exists a unique solution $\hat{u}$ of (4.9) in $W_2^{1,\infty}(\mathbb{R})$, which is twice continuously differentiable and given explicitly as the optimal cost (4.6). Moreover, there exists $\bar{x}$ in $\mathbb{R}$ such that

$$\left.\begin{aligned} &A\hat{u} = f \text{ in }]\bar{x},+\infty[, \\ &\hat{u}'(\bar{x}) = 0 \\ &\hat{u}(x) = \hat{u}(\bar{x}) \text{ in }]-\infty,\bar{x}]. \end{aligned}\right\} \quad (4.11)$$

Furthermore, the following process is optimal,

$$\nu(t) = \nu_1(t) + \int_0^t [\bar{x}-x-\gamma(y(s),\zeta)]^+ d\mu_s(\zeta), \quad (4.12)$$

where $\nu_1(t)$ is the increasing process corresponding to the reflected diffusion with jumps $y(t)$ on $[\bar{x},+\infty[$, with initial conditions $\nu_1(0)=(\bar{x}-x)^+$, $y(0) = \bar{x} \vee x$, and coefficients $g(x)$, $\sigma(x)$, $\bar{\gamma}(x,\zeta) = \bar{x} \vee (x+\gamma(x,\zeta))$.

Outline of the proof

First we show that $\hat{u}$ has the derivative $\hat{u}'$ in the space $W_1^{1,\infty}(\mathbb{R})$, by using the penalized problem

$$Au_\varepsilon + \frac{1}{\varepsilon}(u_\varepsilon')^- = f. \quad (4.13)$$

This equation (4.13) possesses a unique solution and u_ε converges to $\hat{u}$ uniformly on every compact.

Next, by comparison of $\hat{u}$ with u_0 solution of

$$Au_0 = f \tag{4.14}$$

we deduce that $\hat{u}'(r) = 0$ for some r in $\mathbb{R}$. Also, by comparison of $\hat{u}$ with a function of the form

$$w(x) = c(x-R)^+ \ , \ c,R \text{ constants} \tag{4.15}$$

we obtain $\hat{u}'(R) > 0$ if R is large enough.

A main point is to prove that if $\hat{u}'(y) = 0$ then $\hat{u}'(x) = 0$ for every $x \leq y$. To this end, suppose $\hat{u}'(y) = 0$ and that there exists $z < y$ with $\hat{u}'(z) > 0$. Define

$$\bar{z} = \min \{x > z : \hat{u}'(x) = 0 \} \ , \tag{4.16}$$

then

$$\left.\begin{aligned} &A\hat{u} = f \text{ . in } [z, \bar{z}] \ , \\ &\hat{u}''(\bar{z}-) \leq 0. \end{aligned}\right\} \tag{4.17}$$

Hence, setting

$$w(x) = \left\{\begin{aligned} &\hat{u}(x) \text{ if } x \geq \bar{z} \\ &\hat{u}(\bar{z}) \text{ if } x \leq \bar{z} \end{aligned}\right. \tag{4.18}$$

and using the fact that

$$\hat{u}' \geq 0 \text{ implies } Aw(\bar{z}) \leq A\hat{u}(\bar{z}) \leq f(\bar{z}) \tag{4.19}$$

we obtain a subsolution $w(x)$ of the variational inequality (4.9), i.e.

$$Aw \leq f, \ w' \geq 0, \tag{4.20}$$

and $w > \hat{u}$. This is impossible since $\hat{u}$ is the maximum solution of (4.20). Thus, taking

$$\bar{x} = \max \{x : \hat{u}'(x) = 0 \} \tag{4.21}$$

we check that $\bar{x}$ satisfies the requirements of the Theorem.

The verification of the fact that the process (4.12) is optimal follows from the construction of the reflected diffusion with jumps and the Itô's formula for semimartingale (e.g. Meyer [25] and [24]). □

REFERENCES

[1] E.N. Barron and R. Jensen, Optimal Control Problems with No Turning Back, J. Diff. Equations, 36 (1980),pp.223-248.

[2] J.A. Bather and H. Chernoff, Sequential Decisions in the Control of a Spaceship, Proc. Fifth Berkeley Symp. Math. Stat.Prob. Berkeley, Univ. of California Press, Vol.3 (1967),pp.181-207.

[3] V.E. Benes, L.A. Shepp and H.S. Witsenhausen, Some Solvable Stochastic Control Problems, Stochastics,4 (1980),pp.39-83.

[4] A. Bensoussan and J.L. Lions, Applications des Inéquations Variationnelles en Contrôle Stochastique, Dunod, Paris, 1978.

[5] A. Bensoussan and J.L. Lions, Contrôle Impulsionnel et Inéquations Quasi-Variationnelles, Dunod, Paris, 1982.

[6] M.I. Borodowski, A.S. Bratus and F.L. Chernousko, Optimal Impulse Correction Under Random Perturbations, Appl. Math. Mech. (PMM), 39 (1975), pp. 767-775.

[7] A.S. Bratus, Solution of Certain Optimal Correction Problems with Error of Execution of the Control Action, Appl. Math. Mech.(PMM), 38 (1974), pp. 402-408.

[8] F.L. Chernousko, Optimum Correction Under Active Distrubances, Appl. Math. Mech. (PMM), 32 (1968) pp. 196-200.

[9] F.L. Chernousko, Self-Similar Solutions of the Bellman Equation for Optimal Correction of Random Distrubances, Appl. Math. Mech.(PMM), 35 (1971), pp. 291-300.

[10] P.L. Chow, J.L. Menaldi and M. Robin, Additive Control of Stochastic Linear Systems with Finite Horizon, SIAM J. Control Optim., to appear.

[11] W.H. Fleming and R.W. Rishel, Deterministic and Stochastic Optimal Control, Springer-Verlag, New York, 1975.

[12] B. Francis and K. Glover, Bounded Peaking in the Optimal Linear Regular with Cheap Control, IEEE Trans. Automatic Control, AC-23 (1978), pp. 608-617.

[13] J. Gihman and A. Skorohod, Stochastic Differential Equations, Springer-Verlag, Berlin, 1972.

[14] V.K. Gorbunov, Minimax Impulse Correction of Perturbations of a Linear Damped Oscillator, Appl. Math. Mech. (PMM), 40 (1976), pp. 230-237.

[15] J.H. Harrison and A.J. Taylor, Optimal Control of a Brownian Storage System, Stochastic Proc. Appl., 6 (1978), pp. 179-194.

[16] A. Jameson and R.E. O'Malley, Cheap Control of the Time-Invariant Regulor, Appl. Math. Optim., 1 (1975), pp.337-354.

[17] I. Karatzas, The Monotone Follower Problem in Stochastic Decision Theory, Appl. Math. Optim., 7 (1981), pp.175-189.

[18] I. Karatzas, A Class of Singular Stochastic Control Problems, Adv. Appl. Prob., 15 (1983), pp. 225-254.

[19] P. Kokotovic, R.E. O'Malley and P. Sannuti, Singular Perturbations and Order Reduction in Control Theory : An Overview, Automatica, 12 (1976), pp. 123-132.

[20] N.V. Krylov, Controlled Diffusion Processes, Springer-Verlag, New York, 1980.

[21] P.L. Lions and J.L. Menaldi, Optimal Control of Stochastic Integrals and Hamilton-Jacobi-Bellman Equations, Parts I and II, SIAM J. Control Optim., 20 (1982), pp. 58-81 and pp. 82-95.

[22] J.L. Menaldi and M. Robin, On Some Cheap Control Problems for Diffusion Processes, Trans. Am. Math. Soc., 278 (1983), pp. 771-802. See also C.R. Acad. Sc. Paris, Série I, 294 (1982), pp. 541-544.

[23] J.L. Menaldi and M. Robin, On Singular Stochastic Control Problems for Diffusions with Jumps, IEEE Trans. Automatic Control, to appear. See also Proc. 1983 Am. Control Conf., San Francisco, California, June 1983, pp. 1186-1192.

[24] J.L. Menaldi and M. Robin, Reflected Diffusion Processes with Jumps, Ann. Prob., to appear. See also C.R. Acad. Sc; Paris, Serie I, 297 (1983), pp. 533-536.

[25] P.A. Meyer, Cours sur les Intégrales Stochastiques, Lectures Notes in Math., 511 (1976), Springer-Verlag, Berlin, pp. 245-400.

[26] M. Nisio, On a Nonlinear Semigroup Attached to Stochastic Optimal Control, Publ. Res. Inst. Math. Sci., 13 (1976), pp. 513-537.

ON SOME STOCHASTIC OPTIMAL IMPULSE CONTROL PROBLEMS

U. MOSCO

INTRODUCTION

The purpose of this paper is to briefly illustrate some recent results in optimal impulse control theory, in [4],[9],[17].

Let us start by describing some general features of the problems we are dealing with.

We are interested in "systems" which evolve in time according to a stochastic law and are submitted to two kinds of control, so called *impulse control* and *continuous* (or *running*) *control*. Our objective consists in finding and possibly constructing *optimal controls* with respect to some given optimality criterium.

The concept of *impulse control* refers to a situation in which we have to control our system by acting on it discontinuously in time, namely by giving impulses to (the state of) the system at optimally chosen times of intervention (*stopping times*).

The concept of *continuous control* refers to an action on the system which is exerted by operating some control variables which affect the law of evolution of the system between two successive stopping times.

The situation just described arises naturally in many biological, economic or management problems. In particular, the notion of impulse control seems to be particularly adequate in situations in which it is convenient not to perturb the system continuously. That may be the case for instance because the choice of an optimal intervention on the system requires an exogenous decision process which cannot be delegated to some endogenous controlling variables influencing the continuous evolution law itself. That is the case, for example, for several problems in optimal inventory management.

In the following section we shall formulate our control problem more precisely, starting in Sec. 1 with the case of a pure impulse control and considering then in Sec. 2 the more general case of an impulse *and* continuous control.

Our approach is based on dynamic programming methods, leading to certain systems of partial differential inequalities, which we shall describe in Sec. 3.

In Sec. 4 we shall state our existence result of optimal Markov controls and in Sec. 5 we describe how optimal controls can be constructed once the system of p.d.i. has been solved.

1. THE PURE IMPULSE CONTROL PROBLEM

Our main focus here will be on the impulse control and we will disregard for the moment the continuous control.

The dynamics. We suppose that the "system" under consideration has a state $y(s)$ which obeys a stochastic dynamics in the N-dimensional Euclidean space $\mathbb{R}^N$. More precisely, we assume that $y(\cdot)$ is a *stochastic process* with values in a (bounded) open region $\mathcal{O}$ of $\mathbb{R}^N$, obtained by integrating a Itô differential equation

$$(1) \qquad dy(s) = g(y(s),s)ds + \sigma(y(s))dw(s), \quad s \in]0,T[$$

over an underlying probability space (Ω,A,P) with an increasing family of sub σ-algebras F^s of A. The *drift* term of the equation is given by a map $g\colon Q \to \mathbb{R}^N$ $(Q = \mathcal{O} \times]0,T[\,)$, which may be supposed regular enough. The *diffusion* term is given by an N-dimensional Wiener process with a positive definite variance σ, possibly depending on the state. Both $w(\cdot)$ and $y(\cdot)$ are adapted to F^s for any $s \in [0,T]$.

The evolution of the system, starting from an initial time $t \in [0,T]$, is submitted to an impulse control

$$(2) \qquad I = \{\theta^k,\xi^k\}_{k=1,2,\dots}$$

By this we mean that two sequences are given: a nondecreasing sequence of times $\theta^k \in [t,T]$, with $\theta^k \to T$, which are random variables adapted to F^s (*stopping times*) and a sequence of non-negative vectors $\xi^k \in \mathbb{R}^N$, which also are F^s-adapted (*impulses*). In the interval of time $[t,\theta^1[$, the system evolves freely according to (1), from a given initial state

$$(3) \qquad y(t) = x, \quad x \in \mathcal{O};$$

at time θ^1, the free evolution is stopped and the state of the system is made to change (exogenously and instantaneously) from $y((\theta^1)^-)$, which is the limit of $y(s)$ as $s \uparrow \theta^1$, to

$$(3_1) \qquad y(\theta^1) = y((\theta^1)^-) + \xi^1;$$

the state (3_1) is taken as a new initial state and in the interval of time $[\theta^1,\theta^2]$ the evolution is given again by equation (1); at time θ^2, the state is shifted again from $y((\theta^2)^-)$ to $y((\theta^2)^-) + \xi^2$, and so on. In that way a right continuous process

$$y(s) = y_{w,t;I}(s), \quad s \geq t,$$

is defined, which is the trajectory of our system.

A boundary condition on $\partial\mathcal{O}$ is also prescribed, e.g. *absorption*, and the first exit time of $y(\cdot)$ from $\mathcal{O}$ is denoted by τ.

The *cost*. We assume that the optimality criterium is expressed in terms of a cost functional of type

$$(4) \qquad J_{x,t}(I) = E\int_t^{T\wedge\tau} f(y(s),s)ds + \sum_{t\leq\theta^k\leq T\wedge\tau} [c_o + c_1\xi^k]$$

where $y(\cdot) = y_{x,t;I}(\cdot)$ and $T\wedge\tau = \min(T,\tau)$.

This includes a *running cost*, whose density $f\colon Q \to \mathbb{R}$ may be supposed non-negative, and an *impulse cost*, which, for each k, is the sum of a fixed stopping cost $c_o > 0$ and of a cost $c_1\xi$ proportional to the size of the impulse.

An *actualization factor* might be also allowed which has been put=0 above.

The *value function*. The main problem to deal with is that of the existence of an optimal control $\hat{I}$, that is, of an impulse control $\hat{I} = \{\hat{\theta}^k,\hat{\xi}^k\}$ such that

$$(5) \qquad J_{x,t}(\hat{I}) = \inf_I J_{x,t}(I) = :u(x,t)$$

The function u so defined is the so called *value function* or *Hamilton-Jacobi-Bellman* function. It plays a central role in all what follows.

2. THE CONTINUOUS CONTROL

We consider now a more complicated setting for our optimal control problem in which, is addition to an impulse control as described in Sec. 1 also a *continuous* or *running* control is allowed.

Such a continuous control is a process $d(\cdot)$, with values in some region $\mathcal{U}_{ad}$ of the m-dimensional Euclidean space $\mathbb{R}^m$, adapted to the given $\mathcal{F}^s$. The continuous control influences both the dynamics, namely the drift, and the cost, namely the running cost. The function g in the drift term in equation (1) must now be replaced by

$$(6) \qquad g(y(s),s,d(s)) \ ,$$

g being now a map from $\mathcal{O} \times [0,T] \times \mathcal{U}_{ad}$ to $\mathbb{R}^N$.

The cost density f in (4) is now of the type

$$(7) \qquad f(y(s),s,d(s)),$$

f being a (non-negative) function from $\mathcal{O} \times [0,T] \times \mathcal{U}_{ad}$ to $\mathbb{R}$.

Therefore the dynamics will now be described by the equation

$$(8) \qquad dy(s) = g(y(s),s,d(s))ds+\sigma(y(s))d\tilde{w}(s)+\sum_k \xi^k\delta(s-\theta^k), \quad s \in]0,T[$$

with initial condition

$$(9) \qquad y(t) = x$$

and the *cost* functional will have the form

$$(10) \qquad J_{x,t}(d(\cdot),I) = E\{\int_t^T f(y(s),s,d(s))ds + \sum_{t \leq \theta^k \leq T} (c_0 + c_1\xi^k)\},$$

where $y(\cdot)$ is the solution of (8),(9) for given $d(\cdot)$ and I.

As before, an actualization factor could be also allowed, which for simplicity has been put = 0 in (10).

The *value function* is now defined by putting

$$(11) \qquad u(x,t) = \inf J_{x,t}(d(\cdot),I)$$

where $J_{x,t}(\cdot,\cdot)$ is the functional (10) and the infimum is taken over all admissible controls $d(\cdot)$ and I.

Our purpose is to obtain a pair of *optimal* controls $\hat{d}(\cdot)$ and $\hat{I}$: that is, $\hat{d}(\cdot)$ and $\hat{I} = \{\hat{\theta}^k,\hat{\xi}^k\}_{k=1,2,...}$

$$(12) \qquad u(x,t) = J_{x,t}(\hat{d}(\cdot),\hat{I})$$

and possibly give a constructive procedure for determining such $\hat{d}(\cdot)$ and $\hat{I}$.

3. THE DYNAMIC PROGRAMMING APPROACH

The method of *dynamic programming* of R. Bellman has been applied to stochastic control and in particular to stopping time problems by several authors, as W.H. Fleming, N.V. Krylov, A. Bensoussan and J.L. Lions, D.P. Bertsekas, A Friedman, see e.g. [3],[5],[8],[10],[13].

The value function u of the control problem is shown to be the regular solution of certain systems of partial diferential inequalities.

For the impulse control problems stated above, these inequalities were first derived by A. Bensoussan and J.L. Lions ([1],[2]), as *necessary conditions* for optimality. These authors also studied the *sufficiency* of such inequalities, obtaining the first important results in this direction.

The inequalities we are talking about can be written in a strong form as the following *implicit complementarity system*

$$(13)\qquad \begin{cases} u(x,t) \leq Mu(x,t) \text{ for all } (x,t) \in Q \\ -u_t(x,t) + Lu(x,t) \leq H(x,t,\nabla u) \text{ a.e. in } Q \\ (u-M(u))(-u_t + Lu - H(u)) = 0 \text{ a.e. in } Q \end{cases}$$

with suitable additional terminal and lateral conditions on the boundary ∂Q.

The operators L,H and M appearing in (13) can be described as follows.

L is the 2nd order linear uniformly elliptic operator

$$(14)\qquad L = - \sum_{i,j=1}^{n} a_{ij} \frac{\partial^2}{\partial x_i \partial x_j}$$

with coefficients

$$(15)\qquad a_{ij} = \frac{1}{2}(\sigma\cdot\sigma^*)_{ij} \qquad i,j = 1,\ldots,N,$$

arising from the diffusion term of the equation (1). In the special case in which the variance is state independent and $a_{ij} = \delta_{ij}$, then L coincides with the N-dimensional Laplace operator. The operator H in the "pure impulse" control problem of Sec. 1 is the 1st order linear p.d.o.

$$H(x,t,\nabla u) = f + \nabla u \cdot g,$$

where the map g comes from the drift term of the equation (1), and f from the cost functional (4). Above and everywhere else, ∇u denotes the vector $(\frac{\partial u}{\partial x_1},\ldots,\frac{\partial u}{\partial x_n})$.

For the more general problem formulated in Sec. 2, where the continuous control is also taken into account, the operator H is of more complicated type. H is indeed no more a *linear* operator, as before, but it is now a *nonlinear* 1st order p.d.o. in Hamiltonian form, given by

$$(16) \qquad H(x,t,\nabla u) = \inf_{d \in \mathcal{U}_{ad}} \{f(x,t,d) + \nabla u(x,t) \cdot g(x,t,d)\}$$

where f is the density cost function appearing in (10) and g is the function appearing in the drift term of (8).

The operator M, which is typical of impulse control problems, is given by

$$(17) \qquad M(u)(x,t) = c_0 + \inf\{c_1 \xi + u(x+\xi,t) : \xi \geq 0,\ x+\xi \in \bar{\mathcal{O}}\}$$

It is a nonlinear *global* operator. The operator M occurs, in particular, in the definition of the *free-boundary* associated with the system (13): this is the intersection between $Q = \mathcal{O} \times]0,T[$ and the boundary of the region

$$(18) \qquad C = \{(x,t) \in Q : u(x,t) < M(u)(x,t)\}.$$

This free-boundary plays a central role in the construction of an optimal impulse control. Let us also remark that, on the (open) region C, the equation

$$(19) \qquad -u_t(x,t) + Lu(x,t) = H(x,t,\nabla u)$$

is satisfied: C is called the *continuation region*.

4. THE EXISTENCE OF OPTIMAL CONTROLS

Let us now explain how we can show that optimal controls do indeed exist.

For this, let us go back to the system of inequalities (13).

In [1],[2], A. Bensoussan and J.L. Lions proved that, if a solution u of that system exists, *which satisfies the following regularity properties*

$$u \in C^{0}(\bar{Q}), \ -u_t + Lu \in L_2(Q), \tag{20}$$

then such a function u is the Hamilton-Jacobi-Bellman function of the original impulse control problem, that is, u is given by (11): optimal Markov controls $\hat{I}$ exist, which can be constructed, as we already noticed, from the knowledge of the free-boundary of (13) and the values of u and ∇u in the continuation region C, see the following $\hat{d}(\cdot)$ and Sec. 5.

The problem of the existence of optimal controls was thus reduced to that, purely analytical, of the existence of a *regular solution* (20) of the system (13).

In proving the existence of such a regular solution, two main steps can be distinguished. First, one looks for a suitable weak solution of (13) in a Sobolev space and then one proves that such a weak solution has the required regularity (20).

In order to find a weak solution, the system (13), together with the boundary conditions, is written in the form of a *quasi-variational inequality*, that is, a variational inequality with obstacle ψ given implicitely by the map M:

$$\psi := M(u), \tag{21}$$

u being the solution we are looking for.

For u belonging to the natural Sobolev space of the problem, the function ψ given in (21) turns out to be an irregular one in space variables (i.e. $\psi(\cdot,t)$ is not continuous, nor it belongs to Sobolev spaces). Let us remark incidentally here that this example motivated further research in recent times on variational inequalities with *irregular obstacles*, see [9],[21],[22].

The complete proof of the existence of a (maximal) weak solu tion of the quasi-variational inequality of the impulse control was then carried out by several authors at various degree of generality, by combining existence theorems for variational inequalities and suitable fixed point theorems; as (incomplete) references, we mention, for instance, [1],[2],[3],[16],[18],[23],[24].

The study of the regularity properties of the weak solution u is one of the most difficult step of the theory. First, it is proved that u is continuous on $\bar{Q}$(probabilistic methods [1],[2], iterative methods

[11], C^{α}-estimates [4],[9]). This implies, in particular, that the region (18) is open. Then, by exploiting the equation (19) satisfied in the continuation region, the additional regularity of the gradient of u is also established. The main techniques here are the so called *dual estimates* of solutions of variational inequalities as in [9], [12],[16],[19],[20] and *implicit smoothness properties* of the obstacle (21) as in [6], see also [20].

Let us point out that some of the most difficult technical aspects of the theory are originated by the *nonlinearity* of the operator H.

In this regard, the level of the difficulties to be faced depends essentially on the assumption made on the set of admissible controls $\mathcal{U}_{ad}$.

If $\mathcal{U}_{ad}$ is assumed "*a priori*" to be a bounded subset of $\mathbb{R}^m$, then it turns out that $H(x,t,\nabla u)$ has a *linear growth* at ∞ in $|\nabla u|$. In this case, the methods used to solve the "pure" impulse control problem (i.e., linear H) can be adapted by relying on the theory of variational and quasi-variational inequalities involving *monotone operators*. This has been done in [2],[9],[19],[20].

More difficult is the problem when no "a priori" bound is imposed to $\mathcal{U}_{ad}$, e.g. we have $\mathcal{U}_{ad} = \mathbb{R}^m$: this is the case we consider in this paper. The natural assumption on the dependence of g and f on the control variable d are then the following: g is supposed to be *linear* and f *quadratic* in d. These assumptions lead to an Hamiltonian (16) which is now of *quadratic growth* at ∞ in $|\nabla u|$.

All these technical difficulties have been finally overcome and a fairly general (constructive) theorem of existence of optimal controls is now available.

For sake of simplicity we shall state this result in the special case in which the variance matrix σ is state independent, that is

(22) σ is an invertible $N \times N$ constant matrix.

The more general case of a state dependent $\sigma = \sigma(x)$ has also been treated: it requires however some additional Hölderianity assumptions on the density cost f and the drift term g (see [17]).

We assume moreover the following:

(23) $\mathcal{O}$ is a bounded convex open subset of $\mathbb{R}^N$, with a smooth boundary; $0 < T < +\infty$; $\mathcal{U}_{ad}$ is a convex subset of $\mathbb{R}^m$; $N \geq 1$, $m \geq 1$.

Let us notice that also region $\mathcal{O}$ with *corners* will be allowed.

The drift function (6) is assumed to be of the form

(24) $$g(x,t,d) = g_o(x,t,d) + g_1(x,t)\cdot d$$

with $g_o(x,t,d)$ a *continuous* $\mathbb{R}^N$ vector-valued function of $(x,t,d) \subset \bar{\mathcal{O}} \times [0,T] \times \mathcal{U}_{ad}$ and g_1 a *continuous* $N \times N$ matrix-valued function of $(x,t) \in \bar{\mathcal{O}} \times [0,T]$.

The density cost function (7) is supposed to be given by

(25) $$f(x,t,d) = f_o(x,t) + f_1(x,t,d) + f_2(x,t)d\cdot d$$

where $f_o(x,t)$ is *bounded measurable* in $(x,t) \in \mathcal{O} \times]0,T[$ (but control independent), $f_1(x,t,d)$ is *continuous* in $(x,t,d) \in \bar{\mathcal{O}} \times [0,T] \times \mathcal{U}_{ad}$ and $f_2(x,t)$ is a *symmetric positive definite* $m \times m$ matrix which is *continuous* in $(x,t) \in \bar{\mathcal{O}} \times [0,T]$ together with its inverse $f_2^{-1}(x,t)$.

Finally, we assume that the constants c_o, c_1 appearing in the impulse cost in (10) verify

(26) $$c_o > 0 , \quad c_1 \geq 0.$$

Then, we can prove the

THEOREM. *Under the assumptions (22),...(26) there exist a pair of optimal controls* $\hat{d}(\cdot)$ *and* $\hat{I} \equiv \{\hat{\theta}^k, \hat{\xi}^k\}_{k=1,2,...}$ *and this pair is a Markov control.*

The full proof of this theorem was achieved in [17], by extending and completing previous results obtained in [2],[4],[9],[23] for the non-linear problem at hand.

5. THE CONSTRUCTION OF OPTIMAL CONTROLS

As we said in Sec. 4, the proof of the existence of a pair of optimal controls $\hat{d}(\cdot)$ and $\hat{I}$ can be reduced to the proof that there exists a regular solution (20) of the inequalities (13), and the Theorem above was indeed obtained this way.

What we want now to show, following [2], is that *once such a solution* u *has been computed* then we are in a position to follow a "constructive" procedure for the determination of an optimal w pair of optimal controls.

By saying that u has been computed we mean that we know the free-boundary of the continuation region C, (18), and the values of u and of ∇u in C, see Sec. 3.

We can then construct an optimal sequence

$$\hat{\theta}^1,\hat{\xi}^1 \ ; \ \hat{\theta}^2,\hat{\xi}^2 \ ; \ \ldots$$

as follows (we suppose for the moment that no continuous control is applied).

We let the system start its evolution from the initial state x,t: then $y_{x,t}(\cdot)$ is its trajectory obtained by solving the differential equation of the dynamics.

We then put

(27) $\hat{\theta}^1$ = first exit time of $y_{x,t}(\cdot)$ from C.

In order to determine an optimal impulse $\hat{\xi}^1$, we then proceed as follows: We go back to the expression (17) of the implicit obstacle M(u)(x,t), which we now compute for

(28) $$x = y(\hat{\theta}^1-) \quad , \quad t = \hat{\theta}^1$$

(for that we need to now the value $u(y(\theta'-)+\xi,\theta')$ for all $\xi \geq 0$): As an optimal $\hat{\xi}^1$ we then choose any vector $\hat{\xi}'$ which realizes the infimum in M(u)(x,t) for x and t given by (27),(28).

Once the first pair $\hat{\theta}',\hat{\xi}'$ has been determined, the state of the system is (istantaneously) shifted to the state (belonging to C):

$$x = y(\hat{\theta}^1-) + \hat{\xi}^1 \text{ at the time } t = \hat{\theta}^1$$

this is the new initial state from which the evolution of the system in C stats again.

Then $\hat{\theta}^2,\hat{\xi}^2$ are nextly determined with the same procedure as above and so on, recursively, all the optimal sequence $\hat{I} \equiv \{\hat{\theta}^k,\hat{\xi}^k\}_{k=1,2,\ldots}$ will be determined.

As to the continuous control $\hat{d}(\cdot)$, this can be obtained by standard selection procedure, by selecting an optimal $\hat{d}$ realizing the infimum in the expression (16) of the Hamiltonian, where the ∇u is evaluated along the trajectory followed by the state in C.

We refer to [2] for more details.

It should be said, however, that the effective numerical computation of the free-boundary and of the values of u and ∇u in C presents severe difficult, especially for large n.

To these difficulties one has to add those related to the integration of the Itô equation in C, see for instance on this point ref. [14].

To find simplified numerical procedures is one of the main task that future research in this field must accomplish.

CONCLUDING REMARKS

The existence of optimal continuous and impulse controls can be proved under fairly general assumptions on the dynamics and the cost structure.

A constructive procedure leading to optimal controls, is also available.

The effective numerical implementation of the method, however, especially for large state dimensions remains one of the major problem to be confronted with.

REFERENCES

[1] A. BENSOUSSAN, J.L. LIONS: *Contrôle impulsionnel et inéquations quasi-variationnelles d'évolution*, C.R. Acad. Sc. Paris, 276, série A (1973), 1333-1338.

[2] A. BENSOUSSAN, J.L. LIONS: *Optimal Impulse and Continuous Control, Method of Nonlinear Quasi Variational Inequalities*, Trudy Mat. Inst. Steklov, 134, (1975), 5-22.

[3] A. BENSOUSSAN, J.L. LIONS: *Contrôle impulsionnel et inéquations quasi-variationnelles*, Dunod, Paris (1982).

[4] A. BENSOUSSAN, J. FREHSE, U. MOSCO: *A stochastic impulse control problem with quadratic growth Hamiltonian and corresponding quasi-variational inequality*, J. Reine Angew. Math., 331 (1982), 125-145.

[5] D.P. BERTSEKAS, S.E. SHREVE: *Stochastic optimal control: The discrete time case*, Academic Press, New York.

[6] L. CAFFARELLI, A. FRIEDMAN: *Regularity of the solution of the Q.V.I. for the impulse control problem*, Comm. P.D.E., 3 (1978), 745-753.

[7] F. DONATI, M. MATZEU: *On the Strong Solutions of some Nonlinear Evolution Problems in Ordered Banach Spaces*, Boll. UMI, 5, 16-B (1979), 54-73.

[8] W. FLEMING, R. RISHEL: Optimal deterministic and stochastic control, Springer-Verlag, Berlin, 1975.

[9] J. FREHSE, U. MOSCO: *Irregular Obstacle and Quasi-Variational Inequalities of Stochastic Impulse Control*, Ann. Sc. Norm. Sup. Pisa, Serie IV, IX, n. 1 (1982), 105-197.

[10] A. FRIEDMAN: Stochastic differential equations and applications, Vol. 2, Academic Press, New York, 1976.

[11] B. HANOUZET, J.L. JOLY: *Convergence uniforme des itérés definissant la solution d'une inéquation quasi variationnelle abstraite*, C.R. Acad. Sc. Paris, 286, Série A, (1978), 735-738.

[12] J.L. JOLY, U. MOSCO, G.M. TROIANIELLO: *On the regular solution of a quasi-variational inequality connected to a problem of stochastic impulse control*, J. Math. Anal. Appl., 61 (1977), 357-369.

[13] N.V. KRYLOV: Controlled diffusion processes, Springer Verlag, Berlin, 1980.

[14] H. KUSHNER: Probability methods for approximation in stochastic control and elliptic equations, Acad. Press (1977), New York.

[15] O.A. LADYZENSKAJA, V.A. SOLONNIKOV, N.N. URAL'CEVA: *Linear and quasilinear Equations of Parabolic Type*, Transl. of Math. Monographs, 23 (1968).

[16] M. MATZEU, M.A. VIVALDI: *On the regular solution of a nonlinear parabolic quasi-variational inequality related to a stochastic control problem*, Comm. P.D.E. (10), 4, (1979), 1123-1147.

[17] M. MATZEU, U. MOSCO, M.A. VIVALDI: *Sur le problème du contrôle optimal stochastique continu et impulsionnel avec Hamiltonien à croissance quadratique*, C.R. Acad. Sc., Paris, t. 296, Série I (1983), 817-820.

[18] F. MIGNOT, J.P. PUEL: *Inéquations d'évolution paraboliques avec convexes dépendant du temps; applications aux inéquations quasi-variationnelles d'évolution*, Arch. Rat. Mech. An. 64 (1977), 59-91.

[19] U. MOSCO: *Nonlinear quasi-variational inequalities and stochastic impulse control theory*, Proc. Conf. Equadiff IV, Praha, 1977, edited by J. Fabéra, Lect. Notes in Math., 703 , Springer-Verlag, (1979).

[20] U. MOSCO: *On some nonlinear quasi-variational inequalities and implicit complementarity problems in stochastic control theory*, in *Variational Inequalities*, Proc. edited by R.W. Cottle, F. Giannessi and J.L. Lions, J. Wiley.

[21] U. MOSCO: *Module de Wiener et estimations du potential pour le problème d'obstacle*, C.R. Acad. Sci. Paris, to appear.

[22] U. MOSCO: *Obstacle problems : Do continuous solutions exist under wildly irregular constraints?* IIASA Publ., to appear.

[23] M.A. VIVALDI: *A parabolic quasi-variational inequality related to a stochastic impulse control problem with quadratic growth Hamiltonian*, Numer. Funct. Anal. and Optimiz., 4 (3), (1981-82), 241-268.

[24] M.A. VIVALDI: *Non linear parabolic variational inequalities: existence of weak solutions and regularity properties*, to appear.

APPROXIMATION OF HAMILTON-JACOBI-BELLMAN EQUATION IN DETERMINISTIC CONTROL THEORY. AN APPLICATION TO ENERGY PRODUCTION SYSTEMS

Edmundo ROFMAN

Institut National de Recherche en
Informatique et en Automatique
78153 Le Chesnay Cedex
France

INTRODUCTION

Several approximation methods to compute the value function of dynamic optimal control problems can be mentioned in a first presentation.

Following [8] we can divide them in five groups :

I - The vanishing viscosity method
II - Approximation in control space
III - A Trotter formula
IV - Maximizing subsolutions
V - Hyperbolic schemes

In general the practical implementation of these methods involve more or less standard discretisation procedures. It was also the case when the first applications of the "maximizing subsolutions" method were done. The efficiency of this method has recently improved after using a non standard discretisation (cfr. [6]). The aim of this paper is :

a) to present a new result concerning the convergence of that approximation procedure ;

b) to show the impact of several recent contributions in the resolution of the problem presented by González and Rofman in [6] in which the optimization of a small energy production system was discussed.

In § 1 and § 2 we recall the results that are necessary to present the viscosity solution as the limit of the subsolutions and to study a more general short-run model. This model is presented in § 3. The quasi-variational inequalities to be satisfied by the value function are given at the end of this chapter.

In § 4 it is shown that we deal with a sequence of non linear fixed-point problems. They are considered as dynamic programming problems on a graph. Comparative exemples show the advantage of this approach. Finally two problems posed by Electricité de France are solved.

§ 1 THE ORIGINAL PROBLEM AND ITS EQUIVALENT FORMULATION

The system satisfies in absence of impulse controls the differential equation

(1.1) $$\left| \begin{array}{ll} \frac{dy}{ds} = f(y,u,s) & x \in \Omega \subset \mathbb{R}^n \\ y(t) = x & t \in [o,T] \end{array} \right.$$

where u(.) is a measurable function of the time, with values in a compact set $U \subset \mathbb{R}^m$.

In a finite set of times θ_ν $(\nu = 1,2,\ldots\mu)$ impulses $z(\theta_\nu) \in Z$ are applied ; the trajectory jumps are

(1.2) $$y(\theta_\nu^+) = y(\theta_\nu^-) + g(y(\theta_\nu^-),\ z(\theta_\nu),\theta_\nu)$$

Z is a compact set of $\mathbb{R}^p$.

We denote by $(u(.),z(.),\tau)$ a control strategy with the stopping-time $\tau \in [o,T[$.

The cost associated with each strategy is

(1.3) $$J(x,t;u(.),z(.),\tau) = \int_t^{\tau\wedge T} e^{-\alpha(s-t)} \ell(y(s),u(s),s)\, ds + \sum_\nu q(y(\theta_\nu^-),\ z(\theta_\nu),\theta_\nu) e^{-\alpha(\theta_\nu - t)} + e^{\alpha(\tau-t)} \emptyset(y(\tau),\tau) \chi_{[t,T}$$

$\chi_{[t,T[}(.)$ characteristic function of the interval $[t,T[$.

The optimal cost function is

(1.4) $$V(x,t) = \inf\{J(x,t;u(.),z(.),\tau) : u(.),\ z(.),\tau\}\ ,\ \forall (x,t) \in Q$$

(1.5) $$Q = \Omega \times [0,T]$$

In the following we will suppose

i) f, ℓ, ∅, g, q are continuous and bounded functions ; they are lipschitzean functions in (x,t).

ii) $\emptyset(x,T) \geq 0,\ \forall\ x$

iii) $q(x,z,t) \geq q_0 > 0 \quad \forall (x,t) \in Q,\ \forall\ Z$

iv) $\forall\ t,\ y(t) \in \Omega$ independently of the strategy.

We can give the following characterization of $V(x,t)$

<u>Theorem 1.1</u> :

V(x,t) is the maximum element of the set W, with

$W = \{w(x,t) \to \mathbb{R} / (1.6) - (1.10)\}$

(1.6) w(x,t) lipschitzean function in (x,t);

(1.7) $$\frac{\partial w(x,t)}{\partial t} + \min_{u \in U} \left[\frac{\partial w(x,t)}{\partial x} f(x,u,t) + \ell(x,u,t) - \alpha w(x,t)\right] \geq 0$$ a.e. $(x,t) \in Q$;

(1.8) $$w(x,t) \leq \min_{z \in Z} (q(x,z,t) + w(x+g(x,z,t),t))$$ $\forall (x,t) \in Q$;

(1.9) $$w(x,t) \leq \emptyset(x,t)\ ,\ \forall (x,t) \in Q$$

(1.10) $$w(x,T) \leq 0\ ,\ \forall\ x \in \Omega\ .$$

The proof follows the method used in [4], p.29.

§ 2 THE APPROXIMATION PROCEDURE FOR THE OPTIMAL COST

2.1 The discretized problem (P_h)

a) The set Q is approximated with a triangulation Q^h, union of simplices of vertices (x_p, t_p) ; $p = 0, N_x$; $q = 0, N_T$, $t_q = q\delta$, $\delta = \frac{T}{N_T}$

b) In the set of linear finite elements w^h defined in Q^h we consider the set W^h:

$$W^h = \{w^h : Q^h \longrightarrow \mathbb{R} \,/\, (2.1), (2.3), (2.4), (2.5)\}$$

(2.1) $$\frac{\partial w^h}{\partial t}(x_p, t_q; u) + \frac{\partial w^h}{\partial x_f}(x_p, t_q, u) \, \|f(x_p, u, t_q)\| + \ell(x_p, u, t_q) - \alpha\, w^h(x_p, t_q) \geq 0$$

where $\frac{\partial w^h}{\partial x_f}$ is the derivative of w^h in the direction of the vector f and

$$\frac{\partial w^h}{\partial t}(x_p, t_q; u) + \frac{\partial w^h}{\partial x_f}(x_p, t_q ; u) \;\; \|f(x_p, u, t_q)\|$$

is the product of the derivative of w^h in the direction of the vector $(1, f'(x_p, u, t_q))' \in \mathbb{R}^{n+1}$ by the norm of such vector ; for example, in the situation depicted in the following figure

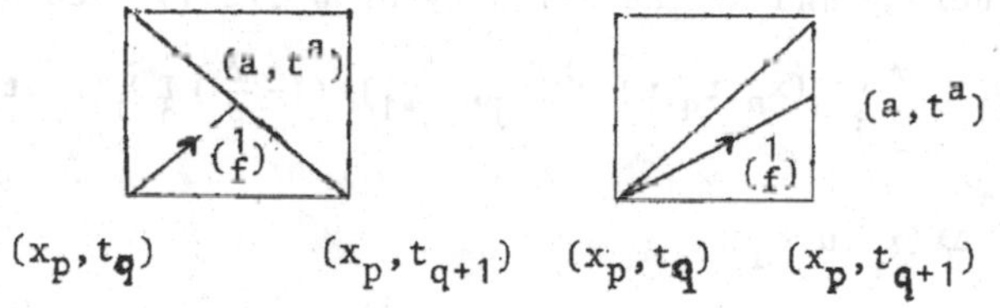

the expression is equal to

(2.2) $$\frac{w^h(a, t^a) - w^h(x_p, t_q)}{\Delta}$$

with $\Delta = t^a - t_q$;

(2.3) $$w^h(x_p,t_q) \le q(x_p,z,t_q) + w^h(x_p+g(x_p,z,t_q),t_q)$$

$$\forall z \in Z^h,\ \forall x_p,\ p = 0, N\ ;\ \forall t_q,\ q = 0, N_T - 1$$

(2.4) $$w^h(x_p,t_q) \le \emptyset\,(x_p,t_q) \quad \forall p = 0,N_x \quad , \forall q = 0, N_T-1$$

(2.5) $$w^h(x_p,t_{N_T}) \le 0 \ , \quad \forall \ p = 0, N_x$$

Remark : The sets U^h, Z^h must satisfy some "consistency conditions" as in ([6] ;2.1,d).

c) We introduce the following partial order "≤"

(2.6) $$w^h \le \tilde{w}^h \iff w^h(x_p,t_q) \le \tilde{w}^h(x_p,t_q)\ ,\ \forall\ p = 0,\ M_x \quad ;\ q = 0,\ M_T$$

and we pose the discretized problem :

(P_h) : Find the maximum element $\bar{w}^{\,h}$ of the set W^h with respect to the partial order " ≤ ".

2.2 <u>The solution of (P_h) and its properties</u>

(2.1) and (2.3) will be transformed into equivalent and more useful relations.

Taking into account that in (2.2), the points (a,t^a) are, in general, interior points of faces (or edges) of some simplex, we will express these points as convex combinations of the vertices of the faces to which they belong and by the affinity of w^h, (2.1) becomes

(2.7) $$w^h(x_p,t_q) \le \min_{u \in U^h} \frac{1}{(1+\alpha\Delta)} \{ \frac{\Delta}{\delta} \sum_j \lambda_j(x_p,t_q,u)\ w^h(x_j,t_{q+1}) + (1-\frac{\Delta}{\delta}) \sum_j \hat{\lambda}_j(x_p,t_q,u)$$
$$w_h(x_j,t_q) + \Delta \ell(x_p,u,t_q)\}$$

in which

(2.8) $$\delta = t_{q+1} - t_q$$

(2.9) $$a = \frac{\Delta}{\delta} (\sum_j \lambda_j . x_j) + (1 - \frac{\Delta}{\delta}) \sum \hat{\lambda}_j . x_j)$$

(2.10) $$\begin{cases}\sum_j \lambda_j(x_p,t_q,u) = 1\\ \sum_j \hat{\lambda}_j(x_p,t_q,u) = 1\end{cases}$$

In the same way we put

(2.11) $$x_p + g(x_p,z,t_q) = \sum_j \lambda'_j(x_p,t_q,z)x_j$$

and (2.3) is rewritten in the equivalent form

(2.12) $$\begin{cases} w^h(x_p,t_q) \leq \min\limits_{z \in Z^h}(q(x_p,z,t_q) + \sum \lambda'_j(x_p,t_q,z)w^h(x_j,t_p))\\ \forall\ p=0,N_x\ ;\ q = 0,N_T-1\end{cases}$$

We will use (2.7) and (2.12) to define the operator M, (w^h denotes a linear finite element in Q^h) :

(2.13) $$\begin{cases} \text{if } q = N_T \quad (Mw^h)(x_p,t_q) = 0\\ \text{if } q = 0,\ldots,N_T-1 \quad (Mw^h)(x_p,t_q) = \min\{\emptyset(x_p,t_q),\\ \quad ,\min\limits_{z \in Z^h}(q(x_p,z,t_q) + \sum_j \lambda'_j(x_p,t_q,z)\, w^h(x_j,t_q)),\\ \quad ,\min\limits_{u \in U^h} \frac{1}{1+\alpha\Delta}[\frac{\Delta}{\delta}\sum_j \lambda_j(x_p,t_q,z)w^h(x_j,t_{q+1}) + (1-\frac{\Delta}{\delta})\\ \quad \sum_j \hat{\lambda}_j(x_p,t_q,u)w^h(x_j,t_q) + \ell(x_p,u,t_q)]\}.\end{cases}$$

We define Mw^h at arbitrary points in Q^h by linear interpolation of the values given by (2.13) at the vertices of the triangulation.

Some properties of Mw^h which follow immediately are :

(2.14) $$w^h \geq \hat{w}^h \Longrightarrow Mw^h \geq M\hat{w}^h$$

(2.15) $$w^h \in W^h \Longleftrightarrow w^h \leq Mw^h$$

Remark : (2.15) give us a characterization of W^h.

Finally the most important property is given by

Theorem 2.1 (see [6])

There exists $\bar{w}^h$, maximum element of W^h; furthermore $\bar{w}^h$ is characterized by the condition $\bar{w}^h = M\bar{w}^h$, i.e.

$$\bar{w}^h = M\bar{w}^h \iff \bar{w}^h \geq w^h, \ \forall w^h \in W^h \tag{2.16}$$

2.3 Algorithm to compute $\bar{w}^h$

To compute $\bar{w}^h$, we can use the following algorithm A1. This algorithm is an improvement of algorithm (I.2.34)[6] because it takes advantage of the particular structure of non-statio nary problems (we use backward solutions) and it computes and approximate solution $\bar{w}_\varepsilon^h$ in a finite number of steps.

Algorithm A1 :

step 0 : choose $\varepsilon > 0$, $\tilde{w}^h \in W^h$ set $w_R^h = \tilde{w}^h$, $\hat{w}^h = \tilde{w}^h$, $w^h = \tilde{w}^h$, $q = N_T$

step 1 : set $p = 0$

step 2 : set $w^h(x_p,t_q) = (Mw^h)(x_p,t_q)$

step 3 : if $p = N_x$, go to step 5 ; else, go to step 4.

step 4 : set $\hat{w}^h(x_p,t_q) = w^h(x_p,t_q)$, and go to step 2 .

step 5 : if $w^h(x_p,t_q) \leq w_R^h(x_p,t_q) + \varepsilon$ for every $p=0,N_x$ go to step 6; else go to step 7.

step 6 : if $q = 0$, set $\bar{w}_\varepsilon^h = w^h$ and stop ; else, set $q=q-1$ and go to step 1.

step 7 : set $w_R^h(x_p,t_q) = w^h(x_p,t_q)$ for every $p = 0,N_x$ and go to step 1.

In fact, it is possible to show the following theorem :

Theorem 2.2 :

The algorithm A1 stops after a finite number of iterations at the element $\bar{w}_\varepsilon^h$, having the following properties :

a) $\bar{w}_\varepsilon^h \in W^h$, $\forall \varepsilon > 0$

b) $\varepsilon \leq \varepsilon' \Longrightarrow \bar{w}_\varepsilon^h \geq \bar{w}_{\varepsilon'}^h$

c) $\lim_{\varepsilon \to 0} \bar{w}_\varepsilon^h = \bar{w}^h$.

2.4 The convergence of the approximate solutions

It is possible to prove a theorem similar to Theorem I.3.2 [6].

Theorem 2.3 :

The approximate solution $\bar{w}^h$ converges uniformly to V(x,t) i.e.

$$\lim_{\|h\| \to 0} \max_{(x,t) \in Q^h} |\bar{w}^h(x,t) - V(x,t)| = 0 \tag{2.17}$$

2.5 Some comments

If assumption i) introduced before Theorem 1.1 is now replaced by

i') f continuous, bounded and lipschitzean function ;
ℓ, $\emptyset$,g,q uniformly continuous and bounded functions in Q,

it is possible to prove (see [9]) that (2.17) holds. Note that in this case the value function V(x,t) is only uniformly continuous and bounded. It can be characterized either as the unique viscosity solution of

$$\frac{\partial V(x,t)}{\partial t} + \min_{u \in U} \frac{\partial V(x,t)}{\partial x} f(x,u,t) + \ell(x,u,t) - \alpha V(x,t)] = 0$$

or as the maximum solution of the set of all uniformly continuous and bounded functions w(x,t) satisfying (1.8),(1.9),(1.10) and, $\forall u \in U$

$$\frac{\partial w(x,t)}{\partial t} + \frac{\partial w(x,t)}{\partial x} f(x,u,t) + \ell(x,u,t) - \alpha w(x,t) \geq 0$$ in the sense of distributions.

The rate of convergence has been discussed in [6] for control policies containing stopping time and impulse controls. In the stationary case and with infinite horizon it is proved the existence of a constant C such that

$$(2.18) \qquad |\bar{w}_h(x) - V(x)| \leq C\,|\log \|h\|\,|\,\sqrt{\|h\|} \qquad \forall x \in \Omega_h$$

holds.

Furthermore, under suitable assumptions and taking advanta ge of a contraction principle the above estimation (2.18) has been improved in some cases. See, i.e. [2], in which continuous and stopping time controls in infinite horizon problems are considered.

§ 3 THE OPTIMAL CONTROL OF AN ENERGY-PRODUCTION SYSTEM

3.1 Modelisation of the problem (short-run model)

The energy production system consists of ν thermopower plants ($P_1, P_2, \ldots, P_\nu$) being their level of production) and μ hydroplants ($y_1, \ldots y_\mu$: hydropower stock ; $\Pi_1, \ldots \Pi_\mu$: hydropower production). D is the demand of electricity and we denote by $P_{\nu+1}$ the production of an additional source, which is available if it is required.

$$(3.1) \qquad D = \sum_{r=1}^{\nu} P_r + \sum_{\ell=1}^{\mu} \Pi_\ell + P_{\nu+1}$$

The cost of the operation is given by

$$(3.2) \qquad J = \int_0^T \Big[\sum_{r=1}^{\nu} c_r P_r(s) + \sum_{\ell=1}^{\mu} c_{h\ell}\,(y_\ell(s))\Pi_\ell(s) + c_{\nu+1}\,P_{\nu+1}\,(s)\Big]\,ds + \sum_{r=1}^{\nu} n_r\,\bar{k}_r$$

n_r is the number of starts of the r thermopower plant, in the interval [0,T].

$\bar{k}_r$ is the cost of each start up.

We suppose c_r, $j=1,\ldots,\nu$ constants and $c_{h\ell}$ are shadow prices obtained after a long-run optimization (about one year). In our problem, we will consider [0,T] one day or one week. (see [7]) .

We will suppose that there are not delays between the start up of termal plant and the instant in which it begins to produce energy. The methodology to be used here can be easily modified to take into account these delays (see [5]).

In this form, the system will be modeled by its internal state (a discrete variable $E = 1,2,3,\dots,2^{\nu}$ showing which thermopower plants are operating) and the continuous variables y_{ℓ} whose evolution equations are

(3.3)
$$\frac{dy}{dt} = A(s) - \Pi(s) \qquad \begin{array}{l} y=(y_1\dots,y_{\mu}) \quad \Pi=(\Pi_1,\dots,\Pi_{\mu}) \\ A=(A_1\dots A_{\mu}) \\ 0 \le y_{\ell} \le y_{\ell,max} \end{array} \qquad \ell = 1,\dots,\mu$$

Where $A_{\ell}(s)$ is the imput of water in the ℓ - plant.

Our aim is to obtain the control strategy giving the minimum of J. The optimal strategy is a decision concerning when the power plants must operate and at what level of production. We look for optimal feed-back policies acting on the instantaneous state (E(s), y (s)) of the system.

3.2 Optimal feed-back policies

Let us consider as parameters the initial hydropower stock $x = (x_1,\dots x_{\mu})$ and the initial time t of the system and let us introduce the optimal cost functions $V_i(x,t)$, $i=1,\dots,2^{\nu}$, $(x,t) \in Q = \prod_{\ell=1}^{\mu}([o,y_{\ell,max}] \times [o,T]$:

(3.4)
$$V_i(x,t) = \inf_{P_1(.),\dots,P_{\nu}(.),\Pi_1(.),\dots,\Pi_{\mu}(.)} J(x,i,t;P_1(.),\dots,P_{\nu}(.),\Pi_1(.)\dots\Pi_{\mu}(.))$$

with J given by (3.2) related to the initial data $(E(t),y(t))=(i,x)$.

In the following we shall note $P(.)=(P_1(.)\dots P_{\nu}(.))$
From $V_i(x,t)$ it will be possible to define the optimal feed-back policies (see [7]).So, our problem is to compute $V_i(x,t)$.
We recall the following :

3.3 Quasi-variational inequalities (QVI) associated with the control problem and characterization of V_i.

It is possible to show (see [4]) that the V_i's are differentiable in a.e $(x,t) \in Q$. Furthermore they verify a.e. in Q (see [4],[7]) the system of Q.V.I. $(i = 1,\dots,2^\nu)$:

(3.5)
$$\frac{\partial V_i}{\partial t} + \min_{(P,\Pi)\in\Gamma_i(x)} \Big(\sum_{\ell=1}^{\mu} \frac{\partial V_i}{\partial x_\ell}(A_\ell - \Pi_\ell) + \sum_{r=1}^{\nu} c_r P_r + \sum_{\ell=1}^{\mu} c_{h\ell}(x)\Pi_\ell + c_{\nu+1}\big(D - \sum_1^{\nu} P_r - \sum_1^{\mu}\Pi_\ell\big)^+ \Big) \geq 0 \; ;$$

(3.6)
$$V_i(x,t) \leq V_j(x,t) + k_j^i \; , \quad \forall j \neq i \; ;$$

(3.7)
$$V_i(x,T) = 0 \; ;$$

(3.8) For a.e. $(x,t) \in Q$ one at least of (3.5) or (3.6) is an equality,

with $\Gamma_i(x)$ the set of admissible levels of production related to the state i and the initial stock x; k_i^j the cost for passing from state i to state j $(j=1,2,\dots,2^\nu)$.

The following characterization of $V_i(x,t)$ will allow us to compute it using the approximation procedure introduced in § 2

$V_i(x,t)$ is the maximum element of the set W_i :

(3.9)
$$W_i = \{ w_i \in H^{1,\infty}(Q) / w_i \text{ verifies (3.5), (3.6), (3.7)} \}$$

i.e. $w_i(x,t) \leq V_i(x,t)$, $\forall \, (x,t) \in Q$, $\forall \, w_i \in W_i$

§ 4 - A FAST METHOD FOR THE SOLUTION OF THE BASIC FIXED POINT PROBLEM

4.1 - The fixed point problem. Some comments and a result of existence and unicity of the solution.

After discretization the system (3.5) to (3.8) gives rise to the problem (2.13), which leads at each time step and each space step to the resolution of a fixed point problem that we may describe in the following form :

With the hypothesis :

(4.1) Let $\phi = \{\ldots,\phi_i,\ldots\}^T \in R^n$; the components ϕ_i of ϕ being all non negatives.

(4.2) Let K be a square $\{n,n\}$ matrix whose coefficients $k_{i,j}$ belong to R. All these coefficients are supposed to be non negative.

We associate to ϕ and K verifying (4.1) - (4.2), the non linear application F , with domain equal to the cone C of vectors of R^n with all components non negative, defined by :

We denote by $a \wedge b$ the minimum of $a,\ b \in R$.

(4.3) Let $w \in C$, then $F(w) \in C$ is such that if $F(w) = \{\ldots,F_i(w),\ldots\}^T$,

$$\forall i \in \{1,\ldots,n\}\ ,\ F_i(w) = \Big(\bigwedge_{j\neq i} (k_{i,j} + w_j)\Big) \wedge \phi_i$$

and we consider the fixed point problem :

(4.4) Find $u \in C$ such that : $u = F(u)$

Problem (4.4) will be in the following denoted as P_1. In this chapter, P_1 will be analyzed as a dynamic programming problem on a graph and a special fast algorithm will be presented to compute the solution. This algorithm, denoted by A_1, use a fixed number $n_{op1} = n \log_2 n$ of operations (additions and comparisons) to compute the solution.

In [10]a "mono-iteration algorithm" A_2 is used to solve P_1. It needs a a number of operations n_{op2} bounded by $n(n-1)/2$. So, A_1 can be faster than A_2 for $n \geq 8$.

Concerning existence and unicity of the solution of P_1 we have the following result :

Proposition 4.1 :

Assume that for all finite sequence of numbers $i_0, i_1, \ldots, i_p$ $(1 \leq i_h \leq n)$

(4.5) $$k_{i_0 i_1} + k_{i_1 i_2} + \ldots + k_{i_p i_0} > 0 \ ;$$

then, there is one and only one solution of P_1

Proof of the unicity

Let be u, $\hat{u}$ two solutions of P_1. After defining $\tilde{u} = u - \hat{u}$ we will prove that $\tilde{u} \equiv 0$.

Let us introduce the sets

(4.6) $$\begin{cases} S = \{i / u_i = \phi_i\} \\ I = \{i / u_i < \phi_i\} = \complement S \ ; \end{cases}$$

so, as $u = F(u)$ we have

(4.7) $$U_i = \bigwedge_{j \neq i} (k_{ij} + u_i)$$

For $i \in I$ we define $\bar{j}(i)$ such that

(4.8) $$u_i = k_{i\bar{j}(i)} + u_{\bar{j}(i)}$$

On the other hand, for $i \in S$, we have

$$u_i = \phi_i \quad , \hat{u}_i \le \phi_i \quad ; \text{ so}$$

(4.9) $$\tilde{u}_i \le 0 \qquad \forall i \in S.$$

We define

$$I_M = \left\{ i \in I / \ \tilde{u}_i = \max_{r \in I} \ \tilde{u}_k \right\}$$

and we prove that it exists $r \in I_M$ such that $\bar{j}(r) \notin I_M$ In fact, if for all $i \in I_M$ we have $\bar{j}(i) \in I_M$ it will be possible to consider a finite set of elements of I_M : $i_0, i_1, \ldots, i_p$ such that

(4.10) $$i_h = \bar{j}(i_{h-1}) \qquad \forall h = 1,\ldots,p$$

$$i_0 = \bar{j}(i_p)$$

Now, from (4.8),

$$u_{i_0} = k_{i_0 i_1} + u_1 = k_{i_0 i_1} + k_{i_1 i_2} + u_2 = \ldots. = k_{i_0 i_1} + \ldots + k_{i_p i_0} + u_{i_0},$$

that is to say, $k_{i_0 i_1} + \ldots + k_{i_p i_0} = 0$, in contradiction with hypothesis (4.5).

Now, we will prove that $\bar{j}(r) \in S$.

In fact, if we suppose $\bar{j}(r) \in I - I_M$, we obtain respectively from (4.3) and (4.8) :

$$\hat{u} \leq k_{r\bar{j}(r)} + \hat{u}_{\bar{j}(r)}$$

$$u_r = k_{r\bar{j}(r)} + u_{\bar{j}(r)} \quad ;$$

in consequence

(4.11) $$\tilde{u}_r \leq \tilde{u}_{\bar{j}(r)}$$

implying

$$\max_{i \in I} \tilde{u}_i = \tilde{u}_r \leq \tilde{u}_{\bar{j}(r)} \leq \max_{i \in I - I_M} u_i < \max_{i \in I} u_i .$$

As this last relation is absurd, $\bar{j}(r) \notin I - I_M$; then $\bar{j}(r) \in S$

Now, using (4.11) and (4.9)

$$\max_{i \in I} \tilde{u}_i = \tilde{u}_r \leq \tilde{u}_{\bar{j}(r)} \leq \max_{i \in S} \tilde{u}_i \leq 0 \ ;$$

so,

(4.12) $$\tilde{u}_i \leq 0 \quad , \forall i$$

In the same way we can prove $\tilde{u}_i \geq 0$, $\forall i$ (because (4.12) uses only that u and $\hat{u}$ are solutions of P_1)

Then

(4.13) $$\hat{u} \equiv u$$

and the unicity of the solution of P_1 is proved.

Proof of the existence

Let us introduce the graph E having n states $i \in (1,2,\dots,n)$ and elementary transitions ij, between them.

A cost k_{ij} is associated to each elementary transition ij. For each state i_0 we consider paths (or concatenations of elementary transitions) $\gamma_{i_0} = (i_0, i_1, \dots, i_p)$.

We denote with p the length of the path and with G_{i_0} the set of paths having i_0 as initial state (including the trivial path given by the unique state i_0, i.e. (i_0)).

To each pair i_0, γ_{i_0} we associate a cost $J(i_0, \gamma_{i_0})$ given by

(4.14) $$J(i_0,\gamma_{i_0}) = \begin{cases} k_{i_0 i_1} + \dots + k_{i_{p-1} i_p} + \phi_{i_p} & \text{if } p \geq 1 \\ \phi_{i_0} & \text{if } p = 0 \end{cases}$$

and we consider the auxiliar problem P_2 :

P_2 : given i_0 find the minimum of the functional $J(i_0, \cdot)$,
i.e. find $V(i_0)$ such that

(4.15) $$V(i_0) = \inf_{G_{i_0}} J(i_0,\gamma_{i_0})$$

To solve P_2 we will use the following

Lemma 4.1 : To search $V(i_0)$ is enough to consider the paths of G_{i_0}

having length $p \leq n-1$.

Proof of lemma 4.1 :

Let be $\gamma_{i_o}^1 = (i_o, i_1, \ldots, i_{p_1})$ a path with length $p_1 \geq n$.

So there is at least one index $\hat{\imath}$ present two times in the path, i.e., there are q, q' such that

(4.16) $$\hat{\imath} = q = q' \qquad q' > q \geq 0$$

The cost of the path $\gamma_{i_o}^1 = (i_o, i_1, \ldots, i_{q-1}, i_q, \ldots, i_{q'}, \ldots, i_{p_1})$ is

$$J(i_o, \gamma_{i_o}^1) = k_{i_o i_1} + \ldots + k_{i_{q-1} i_q} + (k_{i_q i_{q+1}} + \ldots + k_{i_{q'-1}} \; k_{i_{q'}}) +$$

$$+ k_{i_{q'} \, i_{q'+1}} + \ldots + k_{i_{p-1}} \; k_{i_p} + \phi_{i_{p_1}}$$

Because (4.16) and (4.5) we can insure that the path

$$\gamma_{i_o}^2 = (i_o, i_1, \ldots, i_q, i_{q'+1}, \ldots, i_{p_1})$$

has length $p_2 < p_1$ and cost

$$J(i_o, \gamma_{i_o}^2) = k_{i_o} + \ldots + k_{i_{q-1}} + k_{i_{q'} i_{q'+1}} + \ldots + k_{i_{p_1-1} i_{p_1}} + \phi(i_{p_1}) < J(i_o, \gamma_{i_o}^1)$$

In this way we can define a sequence of paths $\gamma_{i_o}^1$, $\gamma_{i_o}^2$, ..., $\gamma_{i_o}^s$ having lengths respectively $p_1 > p_2 > \ldots > p_s$ satisfying $J(i_o, \gamma_{i_o}^s) <$ $< \ldots . < J(i_o, \gamma_{i_o}^2) < J(i_o, \gamma_{i_o}^1)$, and

$$p_s < n$$

So, for the same final state i_{p_1} we have a path $\gamma_{i_o}^{s}$ having length $p_s \leq n-1$ and cost $J(i_o,\gamma_{i_o}^{s}) < J(i_o,\gamma_{i_o}^{1})$. □

So, as the number of paths with length $p \leq n-1$ is finite there exists at least a path $\bar{\gamma}_{i_o}$ with length $\bar{p} \leq n-1$ such that

$$J(i_o,\bar{\gamma}_{i_o}) = \min_{\gamma_{i_o} \in G_{i_o}} J(i_o,\gamma_{i_o}) = V(i_o).$$

So, the function $V(i_o)$ is well defined and has finite values. As the functional J is additive on the path we can obtain, using the dynamic programming approach, that

$$V(i_o) = (\bigwedge_{i \neq i_o} k_{i_o i_1} + V(i_1) \quad \wedge i_o,$$

i.e., the solution of P_1. The existence of solution is now proved and, at the same time, the proposition 4.1 -

4.2 - Structured solution of P_1 : Sequential Dynamic Programming

4.2.1 - Introduction of a stepped decision problem

When P_1 derives from the scheduling problem of a set of thermo generators of electrical energy, it has a special structure that makes possible the solution of P_1 with the use of an algorithm of "sequential dynamic programming type".

In the energy problem above mentioned, the parameter "n", is the number of possible states of a system with "q" thermo generator(*), thus

(*) we will use q at the place of ν in (3.1).

(4.17) $$n = 2^q .$$

We identify each state with a number "i", in such a form that "i" in its binary notation indicates the state of activity of each thermopower plant, with the following convention :

(4.18)
$$\begin{cases} i = \sum_{h=1}^{q} d_h^i \, 2^{h-1} \\ d_h^i = 1 \quad \text{if the } h^{th} \text{ plant is in activity,} \\ d_h^i = 0 \quad \text{if the } h^{th} \text{ plant is inactive.} \end{cases}$$

In this form, we identify the set of "thermic states" with the set of numbers with q -binary digits. Then, we shall use in the following paragraphs, the representation of the state "i" as a vector of q - components :

(4.19) $$i \to d^i = (d_1^i, \ldots, d_q^i).$$

We designate E the set of "thermic states" and we represent it with a graph with (q+1) levels and 2^q points. The level E_h comprises the set of points (or states) that have h-digits with values "1" in their binary notation (h-active plants).

In the graph E, the arrows represent the possible transitions from one level to another (elementary transitions). These transitions take place when there is a start-up of a plant (the state ascends a level) or a shut-down (the state descends a level)

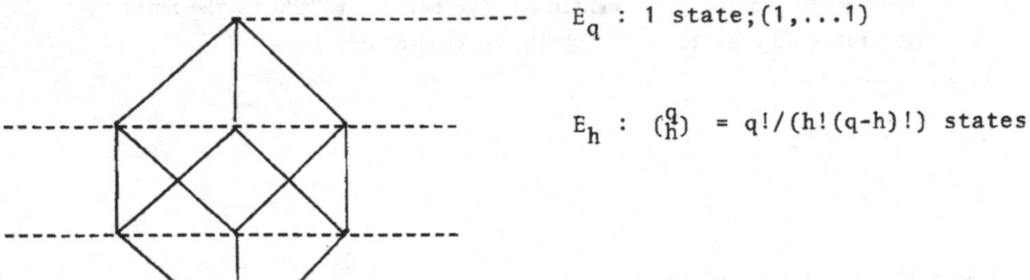

In the energy problem, the parameters k_{ij} are the cost of transition from the state "i" to the state "j". Then, using the notation (4.18) we have :

(4.20) $$k_{ij} = \sum_{h=1}^{q} (d_h^j - d_h^i)^+ . r_h^1 + (d_h^j - d_h^i)^- . r_h^0$$

where

r_h^1 is the start-up cost of plant "h"

r_h^0 is the shut-down cost of plant "h".

The formula (4.20) introduces a cost for the elementary transitions in the graph E, and by concatenation of elementary transitions, for any arbitrary transition.

For each initial state i_0 and to any path

$$\gamma_{i_0} = (i_0, i_1, \ldots, i_p) \ , \ i_h \in \{0, 1, \ldots, 2^q - 1\}$$

being γ_{i_0} a concatenation of elementary transitions, we assign the cost $J(i_0, \gamma_{i_0})$

(4.21) $$J(i_0, \gamma_{i_0}) \begin{cases} = k_{i_0 i_1} + \ldots k_{i_{p-1} i_p} + \phi_{i_p} & \text{if } p \geq 1 \\ = \phi_{i_0} & \text{if } p = 0 \end{cases}$$

We define G_{i_0} as the set of all possible paths γ_{i_0} with origin i_0, and we introduce the following problem :

P_2 : find the function $V(i_0)$, where

(4.22) $$V(i_0) = \inf_{\gamma_{i_0} \in G_{i_0}} J(i_0, \gamma_{i_0})$$

With the usual techniques of dynamic programming theory, it is possible to prove the following property :

Theorem 4.1(*):

Assuming (4.5) satisfied, the function $V(i)$ is well defined and bounded moreover, V is equal to the unique solution u of P_1 and in consequence, P_1 and P_2 are equivalent. □

By virtue of this equivalence, we can solve P_1 applying to P_2 the algorithms given by dynamic programming theory. To develop an efficient algorithm, we shall use the following result:

Theorem 4.2(*):

The optimum of P_2 is realized by a path whose length is $p \leq q$ □

As a consequence, P_2 is equivalent to the following problem P_3, obtained by restriction of P_2 to the set of paths with length $p \leq q$.

P_3 : Find the function $V(i_o)$ defined by :

$$V(i_o) = \min_{\gamma_{i_o} \in G'_{i_o}} J(i_o,\gamma_{i_o}) \tag{4.23}$$

where

$$G'_{i_o} = \{\gamma_{i_o} = (i_o, i_1, .. i_p) / \; p \leq q \;, \; i_h \in \Gamma(i_{h-1}) \;, \; \forall h = 1,p\} \tag{4.24}$$

where

(4.25) $\Gamma(i) = \{j/j$ has at most one binary digit different from those of $i\}$.

4.2.2 Recursive solution of P_3

To solve P_3, we introduce an auxiliary problem P_4.

P_4: Find the function $V^-(i_o)$ defined by

$$V^-(i_o) = \min_{\gamma_{i_o} \in D_{i_o}} J(i_o,\gamma_{i_o}) \tag{4.26}$$

(*) see [3] for the proofs.

where

(4.27) $D_{i_o}=\{\gamma_{i_o}=(i_o,i_1,..,i_p)/\ i_h\in\Gamma^-(i_{h-1}),\ h=1,p\quad p\leq q\}$

and

(4.28) $\Gamma_o^-(i)=\{j/j$ has at most one more binary digit with value "o" than $i\}$

i.e., D_{i_o} is the set of non-ascending paths and $V^-(i_o)$ is the optimum of J in this set of paths.

In the following lines, we give a recursive formula that computes the function $V(i_o)$, using the previously computed values of $V^-(.)$.

For any path, as can be seen from (4.20) and (4.21) the cost J depends only on the digits that change its values and is independent of the order of the changes ; then, any path can be replaced by an equivalent one such that the first part is ascending and the final part is descending.

In consequence, we consider only the paths of the following form :

(4.29)
$$\left|\begin{array}{ll} \gamma_{i_o}=(i_o,i_1,\ldots,i_r,j_1,\ldots,j_s) & \\ s+r\leq q & \\ i_h\in\Gamma^+(i_{h-1}) & \forall h=1,r \\ j_h\in\Gamma_o^-(j_{h-1}) & \forall h=1,s \end{array}\right.$$

(we define $j_o=i_r$)

where

(4.30) $\Gamma^+(i)=\{j/j$ has only one binary digit different from those of i and that digit takes the value "1"$\}$

(4.31) $\Gamma^-(i)=\{j/j$ has only one binary digit different from those of i and that digit takes the value "o"$\}$

(4.32) $\Gamma_o^-(i)=\Gamma^-(i)\cup\{i\}$

We define now two subsets of paths :

(4.33) $D_{i_0} = \{\gamma_{i_0} / r = o\}$

(4.34) $G^+_{i_0} = \{\gamma_{i_0} / r > o\}$

Then, G'_{i_0} is equivalent to $D_{i_0} \cup G^+_{i_0}$ and in consequence :

(4.35) $$V(i_0) = \min_{\gamma_{i_0} \in G'_{i_0}} J(i_0, \gamma_{i_0}) = \min_{\gamma_{i_0} \in D_{i_0} \cup G^+_{i_0}} J(i_0, \gamma_{i_0}) =$$

$$= \min_{\gamma_{i_0} \in G^+_{i_0}} J(i_0, \gamma_{i_0}) \wedge \min_{\gamma_{i_0} \in D_{i_0}} J(i_0, \gamma_{i_0})$$

D_{i_0} is the set of non-ascending paths then we have, by virtue of (5.11)

(4.36) $$V(i_0) = (\min_{\gamma_{i_0} \in G^+_{i_0}} J(i_0, \gamma_{i_0})) \wedge V^-(i_0).$$

To compute the minimum of J in $G^+_{i_0}$, we apply the dynamic programming technique and we obtain the recursive relation :

(4.37) $$\min_{\gamma_{i_0} \in G^+_{i_0}} J(i_0, \gamma_{i_0}) = \min_{i_1 \in \Gamma^+(i_0)} (k_{i_0 i_1} + V(i_1)).$$

Then, from (4.36) - (4.37) we obtain

(4.38) $$V(i_0) = (\min_{i_1 \in \Gamma^+(i_0)} (k_{i_0 i_1} + V(i_1))) \wedge V^-(i_0)$$

Remark 4.1

if $i_0 = 2^q - 1$, it is $\Gamma^+(i_0) = \emptyset$ and then, from (4.38) we have the relation :

$$V(2^q-1) = V^-(2^q-1) \tag{4.39}$$

This relation gives the initial value necessary to start the descending recurrence (4.38) on the graph E, i.e., from (4.38) (4.39) we can compute the values of $V(i_0)$ $\forall\, i_0 \in E_{q-1}$, then the values of $V(i_0)$ $\forall i_0 \in E_{q-2}$, etc., until we reach E_0.

To complete the developement of algorithm A_1 we must obtain a recursive formula to compute $V^-(.)$. In problem P_4, there are two types of possible decisions :

a) the path γ_{i_0} has null length, then

$$J(i_o, \gamma_{i_o}) = \phi_{i_o} \tag{4.40}$$

b) the path γ_{i_0} descends at least one step ; then, applying the dynamic programming principle, we obtain for the minimum with this type of policies, the following value :

$$\min_{i_1 \in \Gamma^-(i_0)} (k_{i_0 i_1} + V^-(i_1)) \tag{4.41}$$

$V^-(i_0)$ is the minimum of the values a, b; then we obtain the following (ascending) recurrence :

$$V^-(i_0) = (\min_{i_1 \in \Gamma^-(i_0)} (k_{i_0 i_1} + V^-(i_1))) \wedge \phi_{i_o} \tag{4.42}$$

Remark 4.2

If $i_0 = 0$, it is $\Gamma^-(0) = \emptyset$; then, from (4.42) we have

$$V^-(o) = \phi_o \tag{4.43}$$

This value is the starting value for the recurrence (4.42) i.e., from (4.42), (4.43) we can obtain all the values $V^-(i_0)$, $\forall i_0 \in E_1$, then $V^-(i_0)$ $\forall i_0 \in E_2$, etc.

4.2.3 The recursive algorithm A_1

step 0 : set $V^-(o) = \phi(o)$, $h = 1$

step 1 : $\forall i \in E_h$,

$$\text{set } V^-(i) = \Big(\bigwedge_{j \in \Gamma^-(i)} (k_{ij} + V^-(j))\Big) \wedge \phi_i$$

step 2 : If $h = q$, go to step 3 ; else

set $h = h+1$ and go to step 1.

step 3 : set $V(2^q-1) = V^-(2^q-1)$, $h = q-1$

step 4 : $\forall\, i \in E_h$, set

$$V(i) = \Big(\bigwedge_{j \in \Gamma^+(i)} k_{ij} + V(j)\Big) \wedge V^-(i)$$

step 5 : if $h = o$, stop ; else, set $h = h-1$ and go to step 4.

4.3 Complexity of algorithm A_1

The complexity of algorithm A_1 is measured by the number of additions or comparisons employed in the computation.

These operations are performed in steps 1 and 4.

In step 1, the total of additions is :

$$n_a^1 = \sum_{h=1}^{q} \binom{q}{h} . h$$

and the number of comparisons is the same :

$$n_c^1 = \sum_{h=1}^{q} \binom{q}{h} . h = n_a^1$$

In step 4, a number of n_a^4 additions and n_c^4 comparisons are performed, where

$$n_a^4 = n_c^4 = \sum_{h=o}^{q-1} \binom{q}{h} (q-h)$$

Then, the total number of additions and comparisons are :

$$n_a = n_a^1 + n_a^4 = \sum_{h=o}^{q} \binom{q}{h} . q = 2^q . q$$

$$n_c = n_c^1 + n_c^4 = q . 2^q$$

From (4.17) we obtain that the complexity of algorithm A1 is given by

.44) $n_{op} = n . \log_2 n$

As the complexity of algorithm A_2 is given by $n_{op} = \frac{1}{2} n(n-1)$, A_1 is faster than A_1 for $n \geq 8$ (corresponding to $q \geq 3$), as it is shown by the following table.

q	n	n(n-1)/2	$n. \log_2 n$
1	2	1	2
2	4	6	8
3	8	28	24

.4 Number of operations and computing times of algorithms A_1 and A_2

We have done the numerical experimentation of both algorithm A_1 and A_2. The computer program makes possible the comparison of the number of operations and the times of computation employed by each algorithm. In the following table, we show the values of times and number of operations obtained for different data (n, ϕ, r_h^o, r_h^1). The values are given for the two computers employed (TI 99 and CII-Honeywell-Bull/DPS 68).

q	n	n_{op_2}	n_{op_1}	TI 99 T_2	TI 99 T_1	CII-HB T_2	CII-HB T_1
3	8	27	24				
4	16	120	64	1'03"	14"	0"05	0"01
4	16	93	64	45"	14"		
4	16	119	64	1'02"	14"		
5	32	491	160	4'30"	37"	0"26	0"02
5	32	386	160	3'17"	37"		
6	64	1838	384	18'22"	1'26"	1"24	0"05
6	64	1491	384	15'30"	1'26"		
6	64	1686	384	16'44"	1'26"		
7	128	6909	896	1h 10'	3'10"		
7	128	8128	896	1h 30'	3'10"	5"79"	0"11
8	256	32641	2048		7'02"	26"13	0"25
9	512	130817	4608			1'56"61	0"67

TABLE 1

The algorithm A_2 obtains the solution in a finite number of operations that use a number of operations $n_{op2} \leq n(n-1)/2$. In the energy problem that is the objectif of this paper, $n=2^q$ and in consequence, n_{op2} takes great values as q grows. The algorithm A_1 (of dynamic programming type) presented in this chapter, also solves P_1 in a fixed number of steps using a number of operations $n_{op1} = n.\log._2 n$ and it is faster than A_2 for $q \geq 3$ $(n \leq 8)$ as is shown in the above table of computing times.

As a final information we give the times needed (on a HB 68) using the graph approach for a whole run of the algorithm during one period of the time interval, with the hydropower stock interval including 19 points of discretization while the thermo power ranges were covered by 5 points :

Number of thermo power plants	2	3	4
Time	1"32	11"	2'09"

TABLE 2

So, rather simple systems can be optimized in real time using this procedure.

4.5 More complex systems (*)

When considering systems with more than one dam we have multidimensional state vectors x in our QVI equations. Discretization schemes are now more complicated and computer times increase significantly each time we add a thermo power plant (Table 3).

Number of dams	2	2	2	2
Number of thermo power plants	2	3	4	5
Time	25"	3'00"	21'00"	$3^h36'$

TABLE 3

Nevertheless, using a more efficient mininization over the power range when solving the QVI'S corresponding to two dams we obtain the times given in table 4.

(*) Particular efforts were necessary in the computing programming task to allow the treatment of more complicated systems. M.C. Bancora-Imbert did this work at INRIA.

Time of HB 68 computer correspond to one period of the time interval when the hydropower ranges and the thermo power ranges were covered by 19 and 5 points respectively.

Number of dams	2	2	2	2
Number of thermo power plants	2	3	4	5
Time	0'5"	0'37"	4'04"	25'00"

TABLE 4

The two following tables give the "sensibility" of the algorithm to variations in the number of discretization points.

Times are always for one period of the time interval. The models considered had two dams.

In table 5 the hydropower range is always covered by 19 points.

In table 6 we retain 4 thermo power discretization points.

Number of thermo power plants \ Number of thermo power discretization points	5	4	3
2	0'5"	4"	3"
3	0'37	24"	15"
4	4'04	2'12"	1'08"
5	25'	12'47"	5'17"

TABLE 5

Number of hydropower discretization points	19	15	11	5
Number of thermo power plants : 4	2'12"	1'22"	44"	13"

TABLE 6

Simulation routines have been written. When typing on a terminal the initial values (i.e. : starting period, number of periods to be simulated, hydraulic stock of different dams, number of thermal plants already working and their identification) the routine prints the desired optimal trajectory to the last period by searching the adequate values in a file previously created during the optimization run. It gives the future state of the system at each period together with the optimal production levels.

4.6 Example 1

The demand is given in table A. Each day is divided in 3 periods. The duration (in hours) of each period is shown in the table.

DEMAND

DAY	MONDAY			TUESDAY			WEDNESDAY			THURSDAY			FRIDAY			SATURDAY			SUNDAY		
PERIOD	1	2	3	1	2	3	1	2	3	1	2	3	1	2	3	1	2	3	1	2	3
Duration (in hours)	10	10	4	8	10	6	8	10	6	8	10	6	8	10	6	4	11	9	4	10	10
Demande (en GW/h)	84.1	78.5	73.0	87.6	83.0	79.5	88.8	84.3	80.9	92.1	87.6	84.3	85.3	80.9	77.9	77.2	68.4	62.8	82.7	75.4	67.8

TABLE A

The energy production systems consists in a dam, several thermo power plants and an additional source which is available if it is required.

The characteristics of the dam are the following :

Maximum capacity of storage	:	7322	GWh
Maximum power (in turbine)	:	6.58	GW
Imputs	:	1.06	GWh
Initial stock	:	287.8	GWh

The cost of KWh in stock is given in c/KWh in table B

DAY	MONDAY	TUESDAY	WEDNESDAY	THURSDAY	FRIDAY	SATURDAY	SUNDAY
Stock between 244.1 and 7322 Gwh	78	74	70	68	65	66	67
Stock between 0 and 244.1 Gwh	160	155	150	148	144	144	145

TABLE B

The admissible level of production in GW, the unit cost in G/KWh and the number of hours of heating (at maximum admissible power) each plant before a start-up are given in TABLE C for the remaining plants :

THERMIC CHARACTERISTICS

DAY	NUCLEAR	COAL 1	COAL 2	FUEL 1	FUEL 2	GAS	ADDITIONAL SOURCE
1 Monday	64,24	5,37	3,62	3,86	1,84	2,11	∞
2 Tuesday	64,61	6,24	3,92	3,65	1,12	2,23	∞
3 Wednesday	66,14	7,70	4,48	4,06	1,66	2,11	∞
4 Thursday	67,09	5,76	3,88	4,12	1,72	2,37	∞
5 Friday	64,27	7,07	3,87	3,44	1,29	2,00	∞
6 Saturday	65,41	7,03	4,80	3,68	0,92	1,86	∞
7 Sunday	66,37	5,57	3,55	3,29	0,85	2,10	∞
Unit cost	5,15	22,0	22.2	50,1	56,5	111,0	200,
Duration of heating (in hours) before a start-up	-	6	6	6	6	-	-

TABLE C

The optimization was done using the graph procedure. For the thermo power plant we consider 4 discretization points :

$$0 \quad ; \quad P_{min} = 0,25 \ P_{max} \quad ; \quad 1/2 \ (P_{min} + P_{max}) \ ; \ P_{max}$$

while the hydropower stock was divided in two intervals.

The time of computation for the whole optimization (the week = 21 periods) was 7"54 on a HB 68 DPS/Multics Computer.

The optimal production policy is shown in the following figures. Nuclear production is omitted (it is always used as the first possibility). The production of COAL 1 is denoted by P_1 ; COAL 2 by P_2 ; FUEL 1 by P_3 ; FUEL 2 by P_4 and GAS by P_5.

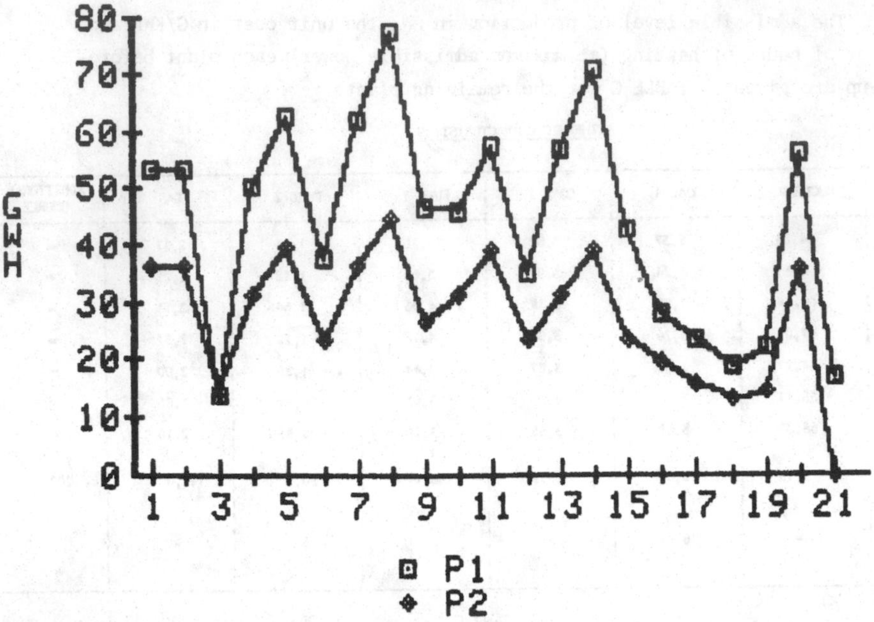
80
70
60
50
40
30
20
10
0
GWH
1
3
5
7
9
11
13
15
17
19
21
P1
P2

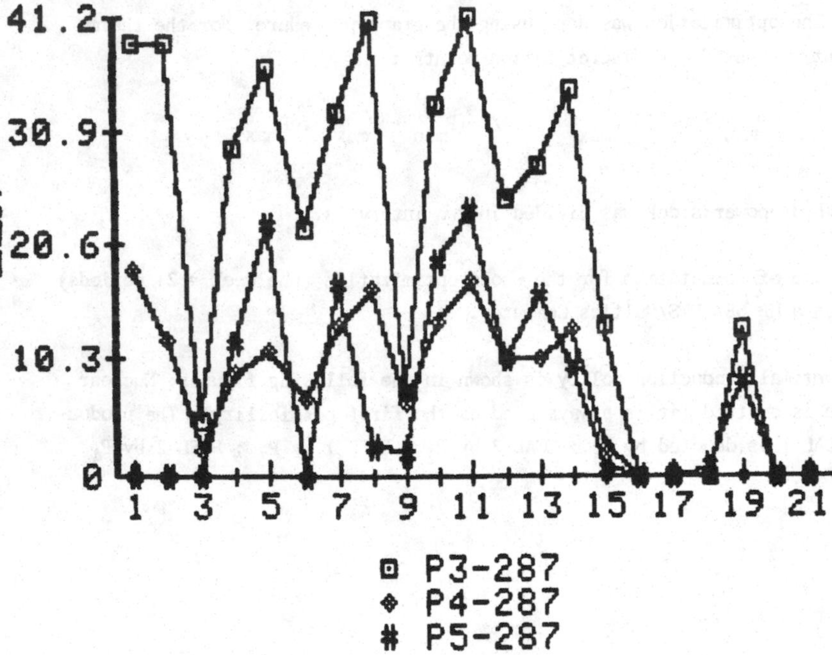
41.2
30.9
20.6
10.3
0
GWH
1
3
5
7
9
11
13
15
17
19
21
P3-287
P4-287
P5-287

THERMO-HYDRAULIC DISTRIBUTION

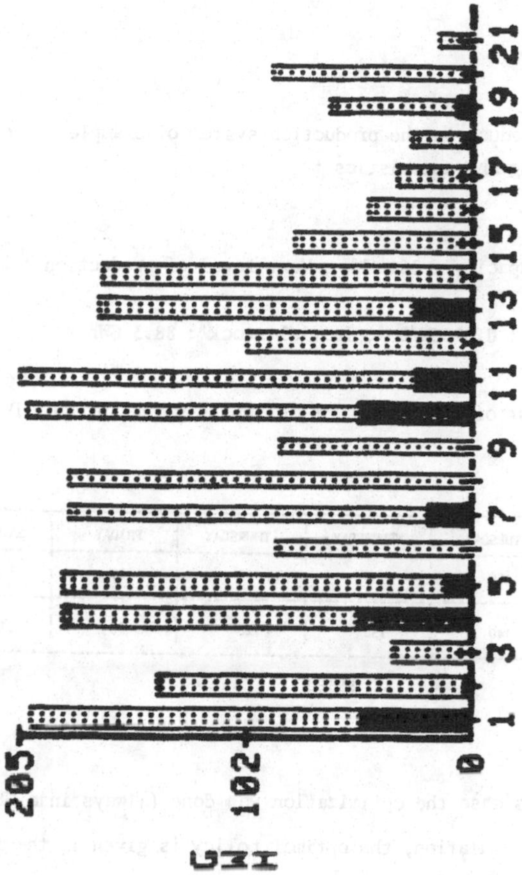

MONDAY-FRIDAY: 3 PERIODS DAILY

EXAMPLE 2

We introduce in the production system of example 1 a second dam with the following characteristics :

Max. Capacity : 361 GWh ; Max. level of production : 2.24 GW

Imputs : 0.01 GW/h ; Initial stock : 88.3 GWh

The cost of production, in c/KWh is given in Table D

DAY	MONDAY	TUESDAY	WEDNESDAY	THURSDAY	FRIDAY	SATURDAY	SUNDAY
Stock between 72.2 and 361 GWh	74	70	65	63	59	60	61
Stock between 0 and 72.2 GWh	145	140	130	120	92	100	110

TABLE D

In this case the optimization was done (always in a HB 68) in 19".

After simulation, the optimal policy is given in the following figures :

2 HYDRAULIC PLANTS

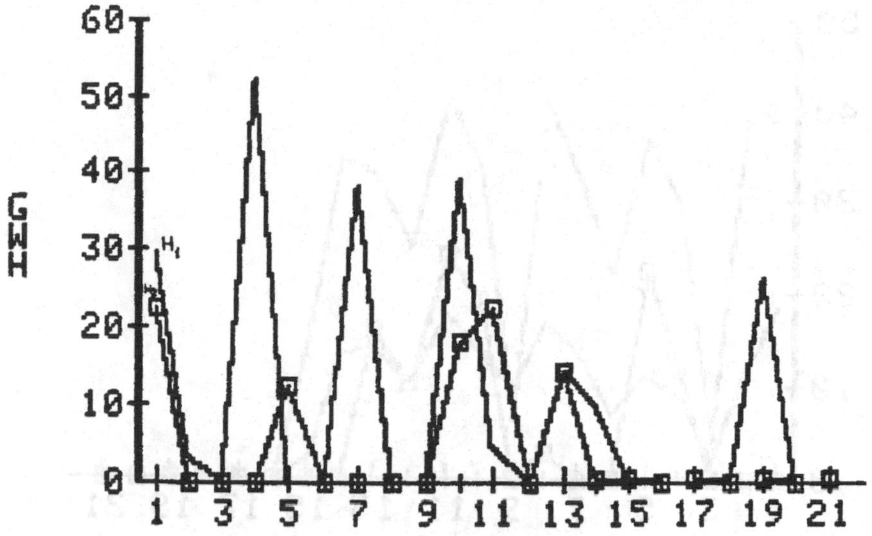

TIME UNITS

COAL POWERED PLANTS (2 HYDR.)

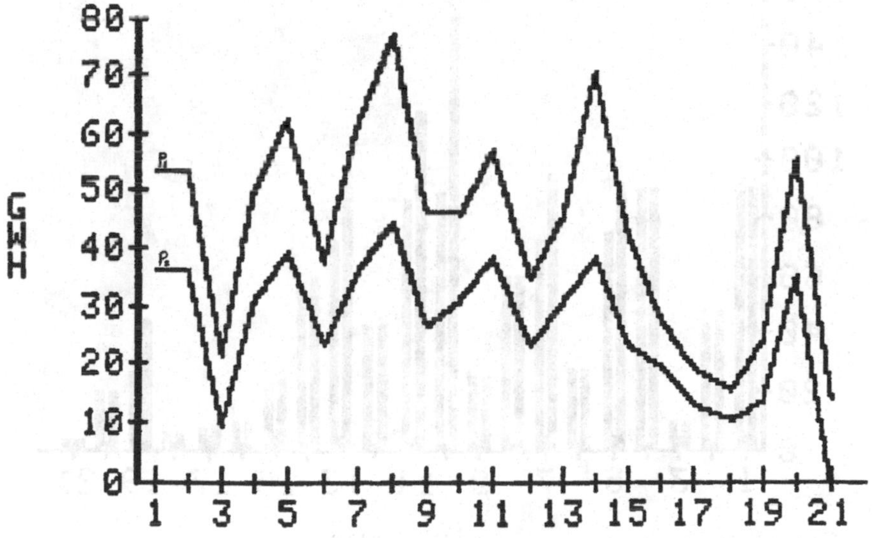

TIME UNITS

FUEL AND GAS POWERED PLANTS (2 HYDR.)

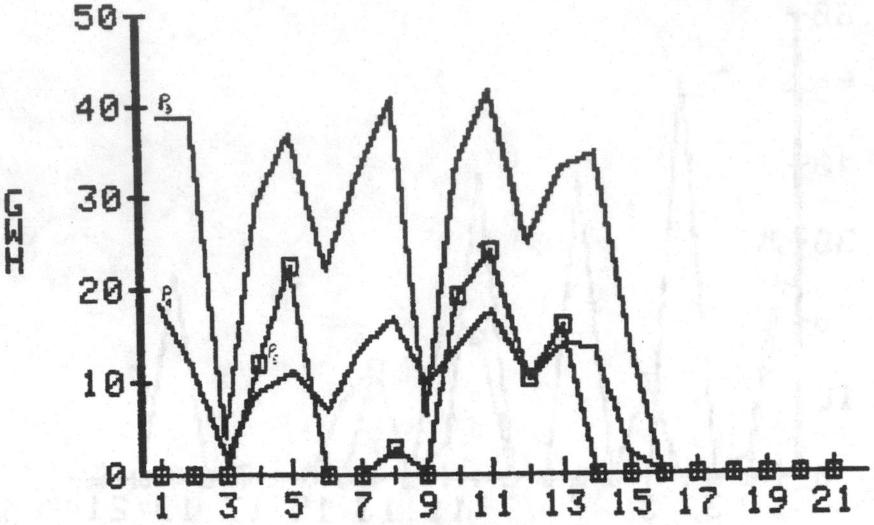

TIME UNITS (GAS:WITH SQUARES)

COST COMPARISON

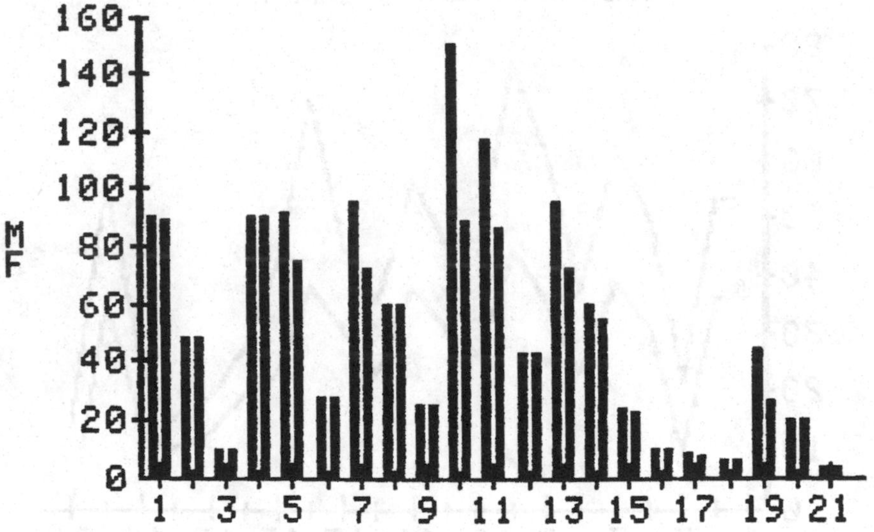

TIME UNITS

COST WITH 1 PLANT: LEFT BAR

COST WITH 2 PLANTS: RIGHT BAR

BIBLIOGRAPHY

[1] A. BENSOUSSAN, J.L. LIONS "Contrôle impulsionnel et inéquations quasi variationnelles", Dunod, Paris 1982.

[2] M. FALCONE "A numerical approach to the infinite horizon problem of deterministic control theory" Preprint Dept. Mat. Univ. Rome, Italie 1984.

[3] R. GONZALEZ "Some contributions to the optimisation of energy production systems" Working paper, Conicet - Univ. Nac. Rosario.

[4] R. GONZALEZ "Sur la résolution de l'éq. de H. Jacobi du Contrôle Déterministe" Cahiers de Math. de la Décision, CEREMADE, n° 8029, 8029 b, 1980.

[5] R. GONZALEZ, E. ROFMAN "Problemas de control bang-bang con intervalos de tiempo minimo entre conmutaciones" in Recents methods in non linear analysis" Ed. De Giorgi-Magenes-Mosco, Pitagora Ed. Bologne 1978.

[6] R. GONZALEZ, E. ROFMAN "On deterministic control problems : an approximation procedure for the optimal cost" part I and II. SIAM J. on Control and Optimization, to appear. See also RR 151, INRIA, Rocquencourt, France.

[7] C. LEGUAY "Application du contrôle stochastique à un problème de gestion optimale d'énergie". Thèse de Docteur-Ingénieur, Université de Paris IX, 1975.

[8] P.L. LIONS "Numerical methods for the H.J.B. equations". Systems and Control Encyclopedia, Pergamon Press Ed, to appear.

[9] J.L. MENALDI, E. ROFMAN, to appear.

[10] J.C. MIELLOU "Sur la résolution itérative d'un problème de point fixe associé à un problème de gestion d'énergie : un algorithm monoiteration" Univ. de Franche-Comté et CNRS E.R.A. de math. n° 070654, Dec. 1982.

DYNAMIC PROGRAMMING FOR OPTIMAL CONTROL PROBLEMS WITH TERMINAL CONSTRAINTS

R.B. Vinter
Department of Electrical Engineering
Imperial College of Science and Technology
London SW7 2BT, UK

Abstract A well-known sufficient condition for optimality is expressed in terms of a continuously differentiable function which is a solution to the Hamilton-Jacobi equation of Dynamic Programming. (A function which serves this purpose is called a Caratheodory function.) However a continuously differentiable may fail to exist, and this limits the usefulness of the condition as classically formulated. Here we ask, how might the condition be modified to extend its applicability? Emphasis is given to problems involving terminal constraints on the trajectories. These pose a special challenge since there is no obvious candidate for a Caratheodory function; we must surmise its existence from abstract arguments, or construct it as the value function of an auxiliary problem. Some interesting connections are made with the theory of necessary conditions.

1. Background

This paper is centered around the following question in optimal control theory. Suppose we have reason to believe a trajectory x is optimal. (Usually x will have been obtained from first order necessary conditions for optimality.) How do we confirm that x is indeed optimal?

The ideas to be described can be presented in a number of guises depending on the way we choose to formulate the optimal control problem. For present purposes it is convenient to adopt the following formulation:

(P) Minimize $g(x(1))$ over Lipschitz continuous functions x such that

$$\dot{x}(t) \in F(t, x(t)), \quad \text{a.e. } t \in [0, 1] \qquad (1.1)$$

$$x(t) \in A \quad , \quad \text{a.e. } t \in [0, 1] \qquad (1.2)$$

$$x(0) = x_0, \; x(1) \in C \, . \qquad (1.3)$$

In problem (P) F is a function with domain $[0, 1] \times \mathbb{R}^n$ which takes values non-empty compact, convex subsets of $\mathbb{R}^n$, g is a real-valued function on $\mathbb{R}^n$. Both A and C are non-empty, closed subsets of $\mathbb{R}^n$. Further conditions will be placed upon the data in problem (P) in due course.

We take a "trajectory" to be an absolutely continuous $\mathbb{R}^n$-valued function x which satisfies the constraints (1.1) - (1.3). A trajectory z is termed "optimal" if the cost functional $x \to g(x(1))$ is minimized at z over the set of trajectories. It is "locally optimal" if there exists some $\varepsilon > 0$ such that the cost function is minimized at z over the set of trajectories x whose graphs lie in the "ε tube" about z, $\{(\tau, \xi): \| \xi - z(\tau) \| < \varepsilon\}$.

There is a well-established methodology for testing whether a trajectory is minimizing, which has been given various names including Dynamic Programming and Caratheodory's Method. (A historical perspective is provided in [6] or [18]). This is expressed in terms of a solution ϕ to the Hamilton-Jacobi equation and associated boundary condition which, for the problem at hand, take the form

$$\left.\begin{aligned} &\frac{\partial \phi}{\partial t}(t, x) - H(t,x, -\frac{\partial \phi}{\partial x}(t, x)) = 0, \quad (t, x) \in [0, 1] \times A \\ &\phi(1, x) = g(x) \quad , \quad x \in C. \end{aligned}\right\} \text{(HJE)}$$

The function H employed here is the hamiltonian:

$$H(t, x, p) := \max_{e \in F(t,x)} p \cdot e.$$

The essential idea of the method is summarized by the following proposition:

<u>Proposition 1.1</u> Let x be a trajectory. Suppose that there exists a continuously differentiable function ϕ defined on some neighbourhood of $[0, 1] \times A$ which satisfies (HJE) on $[0, 1] \times A$ (on some tube about x). Then x is minimizing (locally minimizing) if

$$\phi(0, x_0) = g(x(1)). \tag{1.4}$$

<u>Proof</u> We consider only the global form of the theorem, the local form is proved similarly. Let y be any trajectory. Then since $\phi(1, y(1)) = g(y(1))$ (by the boundary condition) we can write

$$\begin{aligned} \phi(0,x_0) &= \phi(1, y(1) - \int_0^1 \frac{d\phi}{dt}(t, y(t))dt \\ &= g(\quad y(1)) - \int_0^1 [\frac{\partial \phi}{\partial t}(t, y(t)) + \frac{\partial \phi}{\partial x}(t, y(t))\dot{y}(t)]dt \end{aligned}$$

$$\leq g(\quad y(1)) - \int_0^1 [\frac{\partial\phi}{\partial t}(t, y(t)) - H(t, y(t), -\frac{\partial\phi}{\partial x}(t,y(t))]dt$$
$$= g(\quad y(1)).$$

This inequality coupled with (1.4) implies x is minimizing. □

Verification of the minimizing property reduces then to finding a suitable solution to (HJE). Such a function is called a Caratheodory function.

Now a natural candidate for a Caratheodory function is the value function η

$$\eta(t, x) := \inf\{P_{t,x}\}$$

where the right hand side is shorthand for the infimum cost of the optimal control problem ($P_{t,x}$) obtained from (P) by replacing the time interval [0, 1] by [t, 1] and the initial condition x_0 by x. The domain of η comprises points (t, x) such that trajectories exist for problem ($P_{t,x}$). Indeed, when $C = A = \mathbb{R}^n$, and under only mild hypotheses, we find that η is defined on $[0, 1] \times \mathbb{R}^n$, locally Lipschitz continuous on $[0, 1] \times \mathbb{R}^n$ and satisfies (HJE) at all points where it is differentiable (see, e.g. [3]). Knowledge of the value function η leads to information about minimizing trajectories too; if z is optimal and η is continuously differentiable on a tube about x then

$$\frac{\partial\eta}{\partial x}(t, z(t))\cdot\dot{z}(t) = \min_{e\in F(t,x(t))} \frac{\partial\eta}{\partial x}(t, x(t))\cdot e \qquad (1.5)$$

Under favourable circumstances, (1.5) provides a differential equation which, together with the initial condition $x(0) = x_0$, can be solved to yield the optimal trajectory.

These considerations underlie much recent research in Dynamic Programming, research in which the value function plays a central role. Notable questions considered here are, what is a suitable notion of solution to (HJE) (when ϕ is not assumed continuously differentiable), when does such a solution exist, when is it unique and when does it coincide with the value function? (See, e.g., the article by P-L. Lions in this volume). Also, when can x be recovered from equation (1.5)? This is the synthesis problem, which has generated an extensive literature. (See, e.g. [9] or [11]).

Our object here is to report on research of a rather different character. We examine the extent to which the Caratheodory method is applicable when a terminal constraint

$$x(1) \in C$$

is present (and $C \neq \mathbb{R}^n$). The difference is that, for terminally constrained problems, the value function η has at most only a subsidiary role. This is because, unless stringent conditions are imposed on the data, we cannot expect any longer that η will be defined on a sufficiently large subset of $\mathbb{R} \times \mathbb{R}^{n+1}$, or be sufficiently regular, for it to serve as a Caratheodory function. Thus we cannot hope to prove existence of Caratheodory functions by establishing suitable properties of the value function. Instead we must use abstract arguments, or exhibit them as value functions of auxiliary problems which do not involve terminal constraints.

Use of abstract arguments to establish existence of Caratheodory functions for terminally constrained problems was initiated by L.C. Young [18], in the setting of parametric problems in the Calculus of Variations. The basic idea is to interpret Caratheodory functions as subgradients of a certain convex function and then to show, under suitable conditions, that the subdifferential is non-empty. This methodology was applied by Ioffe to fixed endpoint optimal control problems with integral cost, and subsequently refined by Vinter.

Study of particular examples indicates that it is hopeless to try to prove existence of continuously differentiable Caratheodory functions in a general setting. It is natural then to enlarge the family of functions we admit as Caratheodory functions to include non-differentiable ones. It is, of course, required that the Caratheodory functions still retain their basic property of characterizing optimal trajectories. Ioffe [4] introduced a notion of Caratheodory function (also called a Krotov function) which is merely continuous, and proved existence of such a function under very mild conditions.

In a sense however Ioffe's class of Caratheodory functions is too large; the collection of equations and inequalities which replace (HJE) and equ. (1.4) in the new setting are no longer expressed simply in terms of the data of the problem, and are not suitable for generating Caratheodory functions for specific problems. Thus the connection between Caratheodory functions and a usable technique for verifying optimality becomes blurred. Consequently Ioffe proposed consideration of Caratheodory functions which lie in the smaller set of Lipschitz continuous functions. The conditions defining Lipschitz continuous Caratheodory functions <u>are</u> simply expressed in terms of the data, and retain the character of (HJE), etc., of the elementary theory. Ioffe conjectured that Lipschitz continuous

Caratheodory functions exist under "reasonable" assumptions [5].

The class of optimal control problems whose minimizing trajectories are characterized by Lipschitz continuous Caratheodory functions was precisely identified and consequences investigated by Vinter ([12], [13] and [14]). Once again Young's device of proving existence by establishing non-emptiness of a subdifferential was employed. It was subsequently found by Clarke and Vinter ([2], [1]) that the theory concerning existence of Lipschitz continuous Caratheodory functions (at least in a local setting) takes a particularly pleasing form when the optimal control problem is given the formulation (P), and it is possible to exhibit a Lipschitz continuous Caratheodory function as the value function of some auxiliary problem.

It is a remarkable fact that the answer to the question of when do Lipschitz continuous Caratheodory functions exist (in the terminally constrained case) is intimately connected with the theory of necessary conditions: if the necessary conditions in a certain sense give non-trivial information about optimal trajectories (more precisely, if the trajectory in question is "strongly normal") then a Lipschitz continuous Caratheodory function exists which is a generalized solution to the Hamilton-Jacobi equation. This result is unexpected since field theory, which requires us not only to solve the equations supplied by the necessary conditions but also to fit the resulting trajectory into a smoothly parameterized family of "extremals", would seem to suggest that Caratheodory functions can be constructed only under more stringent hypotheses than those required in the theory of necessary conditions.

In the event that the strong normality condition is not satisfied we cannot expect to find Lipschitz continuous Caratheodory functions. As an alternative to considering the continuous Caratheodory functions of Ioffe in this case, we can provide optimality conditions which are closer in spirit to the elementary theory. In these optimality conditions a sequence of continuously differentiable functions replaces the single Caratheodory function however. Results of this nature appear in [15], [16], [17], [7] and [8]. They are proved by interpreting the sequence as a maximizing sequence for the dual problem to some convex optimization problem.

In what follows we summarise salient results from the theory of dynamic programming as it applies to optimal control problems with terminal constraints. In the interests of unity we stick to the formulation (P) throughout, although a variety of formulations are employed in the literature. Consequently certain of the results appear here in a new form.

2. The Strongly Normal Case

The shortcomings of the verification technique implicit in Proposition 1.1, even in the absense of terminal constraints, are illustrated by the following example

Example 2.1 Consider the problem

Minimize $x_2(1)$
subject to

$$\frac{d}{dt}\begin{pmatrix} x_1 \\ x_2 \end{pmatrix}(t) \in \left\{ \begin{pmatrix} 0 \\ x_1 u \end{pmatrix} : u \in [-1, +1] \right\}, \text{ a.e } t \in [0, 1],$$

$$x(0) = 0, \; x(1) \in \mathbb{R}^2 .$$

Here the optimal trajectory is the zero function. However a simple contradiction argument leads to the conclusion that no continuously differentiable Caratheodory function exists confirming optimality.

This simple example is representative of optimal control problems for which extremals can be generated (the necessary conditions are not degenerate), and where we would wish to test extremals for optimality. In seeking a suitable extension of the verification technique which, in particular, embraces Example 2.1, it is natural to consider generalizations of Proposition 1.1 which involve Lipschitz continuous functions. The reasons for this are, firstly, the value function η, which is an obvious candidate for the Caratheodory function in Example 1.1 (by inspection it is $\eta(t,x) = x_2 - |x_1|(1 - t)$ here), is Lipschitz continuous. Secondly, it is possible to interpret Lipschitz continuous functions which are generalized solutions of the Hamilton-Jacobi equation.

Given a point y in an open set $D \subset \mathbb{R}^n$ and a locally Lipschitz continuous function $f: D \to \mathbb{R}$, we denote by $\partial f(y)$ the generalized gradient of f as y, in the sense of Clarke, [1]:

$$\partial f(y) = \{z: z\cdot v \leq \limsup_{\substack{\varepsilon \downarrow 0 \\ y_i \to y}} \frac{f(y_i + \varepsilon v) - f(y_i)}{\varepsilon}, \quad \text{for all } v \in \mathbb{R}^n\}$$

We say that ϕ, a locally Lipschitz continuous function defined on some open subset $\mathcal{D}$ of $]0, 1[\times A$, is a generalized solution of the Hamilton Jacobi equation (on $\mathcal{D}$) if

$$\min_{(\alpha,\beta)\in\partial\phi(t,x)} \{\alpha - H(t, x, -\beta)\} = 0, \quad \text{for all } (t, x) \in D$$

(Notice that generalized solutions ϕ which are continuously differentiable functions are solutions in the classical sense since, in this case, $\partial\phi(t, x)$ is the single point $\{(\partial\phi/\partial t, \partial\phi/\partial x)\}$).

By simple application of the calculus of generalized gradients ([1]) we can show that the assertions of Proposition 1.1 remain true when ϕ is merely a generalized solution of the Hamilton-Jacobi equation. Here is the local version of the result:

<u>Proposition 2.1</u> Let z be an interior trajectory. Suppose that there exists $\delta > 0$ and a Lipschitz continuous function defined on the δ-tube T_δ about z such that

$$\min_{\alpha,\beta\in\partial\phi(t,x)} \{\alpha - H(t, x, -\beta)\} = 0 \quad \text{for all } (t, x) \in \text{interior } \{T_\delta\}, \tag{2.1}$$

$$\phi(1, x) = g(x) \text{ for all } x \in \{\xi: \|\xi - z(1)\| < \delta\} \cap C, \tag{2.2}$$

and

$$\phi(0, x_0) = g(z(1)). \tag{2.3}$$

Then z is locally optimal.

Proposition 2.1 provides confirmation that the zero function is optimal for the problem of Example 2.2; we can take $\phi(t, x) = x_2 - |x_1|\cdot(1-t)$, the value function.

How widely applicable is the verification technique extended in this fashion? To answer this question we must turn to the theory of necessary conditions.

But first we need to narrow the class of problems considered. We shall assume

H1) g is locally Lipschitz continuous on A,

H2) $\text{dist}(F(t', x'), F(t, x)) \to 0$ if $(t', x') \to (t, x)$ in $[0, 1] \times A$,

H3) there exists a constant k such that

$$\text{dist}(F(t, x'), F(t, x)) \leq k|x' - x| \quad \text{for all } (t, x, x') \in [0, 1] \times A \times A$$

H4) there exists a constant r such that

$$|F(t, x)| \leq r \quad \text{for all } (t, x) \in [0, 1] \times A.$$

(Here dist(A, B) denotes the Hausdorff distance of the sets A and B, and $|A|$ denotes $\text{dist}(\{0\}, A)$. We term a trajectory x 'interior' if there exists $\varepsilon > 0$ such that the ε-tube about x is contained in $[0, 1] \times A$.

The following necessary conditions for an interior trajectory to be locally optimal are to be found in [1]:

<u>Theorem 2.1</u> Assume (H1) - (H4). If z is an interior trajectory which is locally optimal, then there exists an absolutely continuous function $p: [0, 1] \to \mathbb{R}^n$ and a number λ equal to 0 or 1 such that

$$(-\dot{p}(t), \dot{z}(t)) \in \partial H(t, z(t), p(t)), \quad \text{a.e. } t \in [0, 1],$$

$$-p(1) \in N_C(z(1)) + \lambda \partial g(z(1)),$$

and

$$\lambda + |p(1)| \text{ is nonzero,}$$

where ∂H refers to the generalized gradient of $(x, p) \to H(t, x, p)$ for each fixed t. $N_C(z(1))$ is the normal cone of C at $z(1)$, [1].

Notice that when λ can be taken equal to 0, then all reference to the cost function g disappears from the conditions. Clearly in such circumstances we cannot expect the necessary conditions to yield useful information about optimal trajectories. Locally optimal trajectories which are not degenerate in this way are termed 'strongly normal':

<u>Definition 2.1</u> An interior, admissible trajectory z which is locally optimal is said to be <u>strongly normal</u> if the only absolutely continuous function p which satisfies

$$(-\dot{p}(t), \dot{z}(t)) \in \partial H(t, x(t), p(t)), \quad -p(1) \in N_C(z(1)),$$

is the zero function.

It follows from Theorem 2.1 that if z is a locally optimal trajectory and $z(1) \in \text{int}\{C\}$ (so that, from a local point of view, the terminal constraint is inoperative) then z is automatically strongly normal, for in this case $N_C = \{0\}$.

And now for the main result of the section. This tells us that, in principle, locally optimal trajectories can always be confirmed as such by means of a local generalized solution to the Hamilton Jacobi equation, when they are interior and strongly normal.

<u>Theorem 2.2</u> Assume (H1) - (H4). Suppose that z is an interior, locally optimal trajectory which is strongly normal. Then there exists $\delta > 0$ and a Lipschitz continuous function ϕ defined on the

δ-tube about z such that (2.1), (2.2) and (2.3) are true.

Proof Consider an auxiliary optimal control problem $(\tilde{P})$ which is the same as (P) except that the terminal constraint "$x(1) \in C$" is now dropped and the cost function is taken to be

$$\tilde{\eta}(x) = k \int_0^1 \max\{\|x - z(t)\| - \varepsilon, 0\}dt + Kd_C(x(1)) + g(x(1)).$$

Here k, K and ε are positive numbers, as yet unspecified, and d_C is the distance function: $d_C(y) = \min\{\|\gamma - y\| : \gamma \in C\}$. It is possible to show that, for k, K and ε suitably chosen, z is an optimal trajectory for $(\tilde{P})$. (Here we make use of the strong normality hypothesis.) Let ϕ be the value function for $(\tilde{P})$. We can also arrange (by choice of k, K, ε), that there exists $\delta \in (0, \varepsilon)$ such that ϕ is defined on a δ tube T_δ about z, is Lipschitz continuous on T_δ and is a generalized solution to the Hamilton Jacobi equation on the interior of T_δ. Thus ϕ satisfies (2.1). It remains to show that ϕ satisfies also equations (2.2) and (2.3). But this is obvious from the fact that ϕ is the value function for $(\tilde{P})$ and z is optimal for $(\tilde{P})$. □

Details of the above proof (which are somewhat intricate) are to be found in [2].

3. General Optimality Conditions

It is possible that optimal trajectories cannot be characterized by Lipschitz continuous Caratheodory functions. The following example demonstrates the point.

Example 1 The problem considered is

$$\begin{cases} \text{Minimize } x_1(1) \\ \text{subject to} \\ \quad \dfrac{d}{dt}\begin{pmatrix} x_1 \\ x_2 \end{pmatrix}(t) \in \left\{ \begin{pmatrix} u \\ 2x_1 u \end{pmatrix} : u \in [-1, +1] \right\}, \text{ a.e. } t \in [0, 1], \\ \begin{pmatrix} x_1 \\ x_2 \end{pmatrix}(0) = 0, \ \begin{pmatrix} x_1 \\ x_2 \end{pmatrix}(1) \in \mathbb{R} \times \{0\}. \end{cases}$$

The zero function is the only trajectory, and this is therefore the optimal trajectory. It can be shown however that no Lipschitz continuous Caratheodory function exists confirming this. There is no contradiction of Theorem 2.2 here for the optimal trajectory is not strongly normal, as is easily checked.

We conclude by presenting results which are illustrative of how dynamic programming ideas can provide optimality conditions even in these circumstances. The hypotheses on the data we shall impose are

$\overline{H1}$) g is continuous on A,

$\overline{H2}$) $\text{dist}(F(t', x'), F(t, x)) \to 0$ if $(t', x') \to (t, x)$ in $[0, 1] \times A$,

$\overline{H3}$) A is compact,

$\overline{H4}$) There exists a constant r such that

$$|F(t, x)| \leq r \text{ for all } (t, x) \in [0, 1] \times A.$$

Note that by introducing hypothesis ($\overline{H3}$) we do not exclude such problems as that treated in Example 2.1. Here the "state constraint" set A is purely notional, and is chosen, say, to contain the values of all trajectories.

We need also to introduce the "reachable set" R:

$$R = \{x(1): x \text{ is a trajectory}\}.$$

Theorem 3.1 Assume ($\overline{H1}$) - ($\overline{H4}$), and suppose that trajectories exist. Then there exists a neighbourhood $\mathcal{D}$ of $[0, 1] \times A$ and a sequence $\{\phi^i\}$ in $C^1(\mathcal{D})$ satisfying

$$\frac{\partial \phi^i}{\partial t}(t, x) - H(t, x, -\frac{\partial \phi^i}{\partial x}(t, x)) \geq 0, \text{ for all } (t, x) \in [0, 1] \times A,$$

$$\phi^i(1, x) \leq g(x) \quad , \text{ for all } x \in C \cap R,$$

along the sequence, such that

a trajectory z is optimal if and only if

$$\phi^i(0, x_0) \to g(z(1)).$$

Implicit in Theorem 3.1 is an extension of the Caratheodory method in which a sequence $\{\phi^i\}$ replaces the single Caratheodory function of Proposition 1.1, and both the Hamilton-Jacobi equation and boundary condition are relaxed to inequalities. It is a simple exercise to show that we retain a sufficient condition for optimality even after making these modifications. The main assertion of Theorem 3.1 is that the condition is also necessary.

Proof The idea of the proof is to replace (P) by an equivalent, convex problem, to which we apply techniques of convex analysis.

First we make a few observations on the notation and terminology to be employed. Let S be a compact set. The norm on C(S) is taken to be the sup norm and that on $C^*(S)$ (the topological dual of $C(S)$) is the usual dual norm. The action of an element μ in $C^*(S)$ on elements m in $C(S)$ is denoted by $m \to \int m d\mu$. Thus we do not distinguish between elements in $C^*(S)$ and the measures which represent them. We write $m \geq 0$ if the measure is non-negative.

Consider now the minimization problem (W):

$$\left\{\begin{array}{l} \text{Minimize } \int g d\beta \text{ over } (\mu, \beta) \in C^*([0,1] \times A \times B) \times C^*(R \cap C) \\ \text{such that} \\ \quad \int (\phi_t(t,x) + \phi_x(t,x)v)\, d\mu(t,x,v) = \int \phi(1,x) d\beta - \phi(0,x_0) \int d\beta \\ \qquad\qquad \text{for all } \phi \in C^1(\mathcal{D}) \qquad (3.1) \\ \quad \int \chi_{\{(t,x,v): v \notin F(t,x)\}} d\mu = 0\,, \\ \quad \mu \geq 0\,, \\ \quad \beta \geq 0 \text{ and } \|\beta\| \leq 1. \text{ (B is the closed ball in } \mathbb{R}^n\text{, radius r)} \end{array}\right.$$

The relationship between (P) and (W) is that feasible elements for problem (P) (i.e. trajectories) can be embedded in the set of feasible elements for problem (W) and, when we pass from (P) to (W), the value of the cost is unaltered. To be more specific, given a trajectory x, we associate with it $\mu (\in C^*([0,1] \times A \times B))$ and $\beta (\in C^*(C \cap R))$ defined by

$$\int m d\mu = \int_0^1 m(t, x(t), \dot{x}(t)) dt \text{ for all } m \in C([0,1] \times A \times B)$$

and

$$\int b d\beta = b(x(1)), \text{ for all } b \in C(C \cap R).$$

The elements μ and β do indeed satisfy the constraints of problem (W). Notice, in particular, that for any $\phi \in C^1(\mathcal{D})$

$$\int (\phi_t(t,x) + \phi_x(t,x)v) d\mu(t,x,v) = \int \frac{d}{dt}\phi(t, x(t) dt$$

$$= \phi(1, x(1)) - \phi(0, x_0) = \int \phi(1,x) d\beta - \phi(0, x_0) \int d\beta,$$

as required by constraint (3.1).

Actually there are feasible elements for (W) which do not arise from trajectories. But (P) and (W) can be shown to be equivalent, to the extent that the minimum in (W) is achieved at an embedded trajectory. The arguments involved are similar to those used in [15].

Next we follow a by now standard procedure to generate a dual problem to (W). The problem (W) is embedded in a family of problems in which the constraint "$\mu \geq 0$" is replaced by "$\mu + e \geq 0$". The perturbation parameter e ranges over $C^*([0, 1] \times A \times B)$. It is a fairly routine matter to show that this family of perturbed problems induces a dual maximization problem, as described in [10], whose supremum cost coincides with that of the maximization problem (D):

$$(D) \begin{cases} \text{Maximize } \phi(0, x_0) \\ \text{over } \phi \in C^1(\mathcal{D}) \text{ such that} \\ \frac{\partial\phi}{\partial t}(t, x) - H(t, x, -\frac{\partial\phi}{\partial x}(t, x)) \geq 0 \\ \qquad\qquad \text{for all } (t, x) \in [0, 1] \times A \\ \text{and} \\ \phi(1, x) \leq g(x), \quad \text{for all } x \in \mathcal{R} \cap C. \end{cases}$$

It is not difficult to show (c.f. [16]) that, although (D) does not necessarily have a solution, the supremum cost for (D) coincides with the minimum cost for (W) (and therefore (P) by equivalence). In the language of convex optimization theory, "strong duality" prevails. The assertions of Theorem 3.1 will now be recognized as an expanded statement of this property. (The sequence $\{\phi^i\}$ is a maximizing sequence for (D)). □

References

[1] F.H. Clarke. Optimization and Nonsmooth Analysis. Wiley, New York, 1983.

[2] F.H. Clarke and R.B. Vinter. Local Optimality Conditions and Lipschitzian Solutions to the Hamilton-Jacobi Equation. SIAM J. Control and Opt., 21, 6, 1983.

[3] W.H. Fleming and R.W. Rishel. Deterministic and Stochastic Optimal Control. Springer, New York, 1975.

[4] A.D. Ioffe. Convex Functions Occurring in Variational Problems and the Absolute Minimum Problem. Mat. Sb., 88, 1972 and Math. USSR - Sb., 1972.

[5] A.D. Ioffe, Private Communication.

[6] A.D. Ioffe and V.M. Tihomirov. Theory of Extremal Problems. North Holland, Amsterdam, 1979.

[7] R.M. Lewis and R.B. Vinter. New Representation Theorems for Consistent Flows, Proc. London Math. Soc., 3, 4, 1980.

[8] R.M. Lewis and R.B. Vinter. Relaxation of Optimal Control Problems to Equivalent Convex Programs. J. Math. An. and Appl., 74, 2,1980.

[9] S. Mirica. Dynamic Programming for Stratified Optimal Control Problems. To appear.

[10] R.T. Rockafellar. Conjugate Duality and Optimization. Regional Conference Series in Applied Mathematics, SIAM, Philadelphia, 1974.

[11] H. Sussman. Subanalytic Sets and Feedback Control. J. Diff. Equs., 31, 1979.

[12] R.B. Vinter. Weakest Conditions for Existence of Lipschitz Continuous Krotov Functions in Optimal Control Theory. SIAM J. Control and Opt., 21, 2, 1983.

[13] R.B. Vinter. New Global Optimality Conditions in Optimal Control Theory, SIAM J. Control and Opt., 21, 2, 1983.

[14] R.B. Vinter. The Equivalence of "Strong Calmness" and "Calmness" in Optimal Control Theory. J. Math. An. and Appl., 91, 1, 1983.

[15] R.B. Vinter and R.M. Lewis. The Equivalence of Strong and Weak Formulations for Certain Problems in Optimal Control. SIAM J. Control and Opt., 16, 4, 1978.

[16] R.B. Vinter. A Necessary and Sufficient Condition for Optimality of Dynamic Programming Type, Making No A Priori Assumptions on the Controls. SIAM J. Control and Opt., 16, 4, 1978.

[17] R.B. Vinter. A Verification Theorem which Provides a Necessary and Sufficient Condition for Optimality. IEEE Trans. Aut. Control, 25, 1, 1080.

[18] L.C. Young. Lectures in the Calculus of Variations and Optimal Control Theory. Saunders, Philadelphia, 1969.